MANAGEMENT
FOR
ENGINEERS

PRENTICE-HALL INTERNATIONAL SERIES
IN INDUSTRIAL AND SYSTEMS ENGINEERING

W.J. Fabrycky and J.H. Mize, Editors

AMOS AND SARCHET *Management for Engineers*
BEIGHTLER, PHILLIPS, AND WILDE *Fundamentals of Optimization, 2nd ed.*
BLANCHARD *Logistics Engineering and Management, 2nd ed.*
BLANCHARD AND FABRYCKY *Systems Engineering and Analysis*
BROWN *Systems Analysis and Design for Safety*
BUSSEY *The Economic Analysis of Industrial Projects*
FABRYCKY, GHARE, AND TORGERSEN *Industrial Operations Research*
FRANCIS AND WHITE *Facility Layout and Location: An Analytical Approach*
GOTTFRIED AND WEISMAN *Introduction to Optimization Theory*
HAMMER *Product Safety Management and Engineering*
IGNIZIO *Linear Programming in Single & Multiobjective Systems*
KIRKPATRICK *Introductory Statistics and Probability for Engineering, Science, and Technology*
MIZE, WHITE, AND BROOKS *Operations Planning and Control*
MUNDEL *Motion and Time Study: Improving Productivity, 5th ed.*
OSTWALD *Cost Estimating for Engineering and Management*
PHILLIPS AND GARCIA-DIAZ *Fundamentals of Network Analysis*
SIVAZLIAN AND STANFEL *Analysis of Systems in Operations Research*
SIVAZLIAN AND STANFEL *Optimization Techniques in Operations Research*
THUESEN, FABRYCKY, AND THUESEN *Engineering Economy, 5th ed.*
TURNER, MIZE, AND CASE *Introduction to Industrial and Systems Engineering*
WHITEHOUSE *Systems Analysis and Design Using Network Techniques*

MANAGEMENT
FOR
ENGINEERS

JOHN M. AMOS

Professor

BERNARD R. SARCHET

Robert B. Koplar Professor
and Founding Chairman

Department of Engineering Management
University of Missouri-Rolla

PRENTICE-HALL, INC.
Englewood Cliffs, New Jersey 07632

Library of Congress Cataloging in Publication Data

Amos, John M
  Management for engineers.

  Includes bibliographies and index.
  1. Engineering–Management.   I. Sarchet, Bernard R.
(date),   joint author.   II. Title.
TA190.A56   620′.0068   80-28391
ISBN   0-13-549402-8

To Ruth and Le

Editorial/production supervision
and interior design by *Virginia Huebner*
Manufacturing buyer   *Joyce Levatino*

© 1981 by Prentice-Hall, Inc.,
Englewood Cliffs, N.J. 07632

Printed in the United States of America

10   9   8   7   6   5   4   3   2

Prentice-Hall International, Inc., *London*
Prentice-Hall of Australia Pty. Limited, *Sydney*
Prentice-Hall of Canada, Ltd., *Toronto*
Prentice-Hall of India Private Limited, *New Delhi*
Prentice-Hall of Japan, Inc., *Tokyo*
Prentice-Hall of Southeast Asia Pte. Ltd., *Singapore*
Whitehall Books Limited, Wellington, *New Zealand*

CONTENTS

LIST OF TABLES

PREFACE

No one who attempts to depict the spirit of the age in which we live can possibly overlook the importance of engineering in today's culture. It assumes a commanding influence on the thoughts of society and, in turn, its organizations and their leaders. We see the evidence all around, whether or not we approve of the trend.

The manager in an industrial enterprise must now be concerned with products or services that are technical in nature. He* is faced with the need to find solutions using engineering techniques.

The demand for engineering education in colleges and universities has reached unprecedented heights. Already more than half of the chief executives of major United States corporations are engineers. In coming years, most of the middle management positions will be filled by engineers, placing society's future literally in the hands of people educated in this discipline.

It is important, therefore, to understand the temper of these individuals, to clarify their aims and to discover their means for achieving them. To understand

---

* The pronoun "he" is used for the third person singular in this book, but is intended to mean both male and female.

the engineer effectively, one must consider the growth and development of his personality.

The foundation for this book was laid in 1967 with the beginning of the Department of Engineering Management at the University of Missouri-Rolla. Two things were apparent at that time. One was the need for a Bachelor of Science program that would provide a graduate with approximately three years of engineering and one year of management, so that he could perform more effectively in the production and marketing areas of the industrial corporation.

The second was a program at the Master of Science level to equip an engineering graduate with the tools of management so that he might become a better engineer. More than 1,600 students have graduated with bachelor's and master's degrees in Engineering Management.

Recent surveys (refer to Sarchet, 1977[1]; Kocaoglu[2]; and Easter and Sarchet, 1980[3]) indicate that interest in this type of education is rapidly increasing, with 87 institutions now offering bachelor's and/or master's degrees. It is anticipated that by 1985 more than 100 institutions will be offering degrees of this type.

This book is of value in three areas: first, for a course at the master's level in engineering management or traditional engineering master's degrees. Engineers who have completed four years of a traditional engineering curriculum frequently learn from interviewers that their education is deficient in an important area, i.e., the area of management. As a result, many stay on to learn the tools of management before they enter their profession. Many others, however, who graduated before such programs were available, now find themselves, ten or twenty years later, either blocked from further promotion because of the lack of management education or trapped in a job where a management education is needed. A course based on this book is useful in either of these situations.

Second, the book is of value as a senior elective course in either traditional engineering or engineering management undergraduate programs, and third, for short courses in management offered to the engineer in a supervisory position who can neither leave his job nor is conveniently located to attend a night master's program. Such instruction will assist him in combatting the problems that he will face as he undertakes the supervisory function. It will help him to use his time and talents effectively.

The book is divided into four parts as described below:

Part I — The Transition of the Engineer to an Engineer Manager. This part consists of one chapter dealing with the change in environment faced by the engineer who becomes a manager, and a second chapter dealing with the decision-making responsibilities he will be assuming.

Part II — Performance of Managerial Functions. This part deals with matters related to the planning, organizing, staffing, directing, and controlling of engineering organizations. The student will be introduced to such techniques as the matrix system of management, motivation, appraisal systems, participative management, and controls.

Part III — Techniques of Engineering Management. This part consists of three chapters dealing with effective ways of presenting engineering ideas, designing and using information systems, and efficient time management.

Part IV — Engineering Management Tomorrow. Some of the problems the engineer manager will face in the future along with some keys to success as engineer or engineer manager are presented. Also included are three separate sections on law (environment, regulations, and patents) and one on ethics.

Each chapter has learning objectives, incidents for discussion, important terms, questions for discussion, and references for further reading.

A unique feature of this book is provided through an introduction to each chapter written by engineer managers from major corporations. In these brief writings, they present their views on the chapter subject matter. This helps to focus the students' thoughts on the material presented in the chapter.

Ours is a technical society, in which engineers have a great influence on the course that we will follow. Under the leadership of qualified engineer managers we can be certain that the specialized talents of the engineers will be used most effectively for the general good of society.

Acknowledgements

A book is always the product of many people. This one is no exception. We owe our thanks to students upon whom the product was tested and to our colleagues in the department. Special appreciation is expressed to Mrs. Paula Tochen Bell, a graduate student who patiently worked with us in reviewing student comments, editing and monitoring the finished product. And to Dr. Dale Jackson of the University of Missouri-Rolla (UMR) for his assistance on the content of several chapters and to Dr. Henry Sineath and Dr. Daniel Babcock of UMR for their critical review.

We also want to thank our secretaries, Mrs. Peggy Pyron, Mrs. Donna Kreisler, Mrs. Irene Sherman, Mrs. Gerre McKay, Mrs. Glenna Grisham, Mrs. Mary Uhlmansiek and Mrs. Betty Lindsey. Without their steadfast work through several drafts, this book would never have become a reality.

JOHN M. AMOS
BERNARD R. SARCHET

Rolla, Missouri

THE TRANSITION OF THE ENGINEER TO AN ENGINEER MANAGER

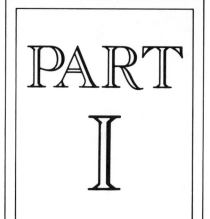

PART

I

THE ENGINEER AS AN ENGINEER MANAGER

The work of engineers is technical in nature and affects large segments of our society. The transition of an engineer to a managerial role requires change in activities, change in attitudes and viewpoints toward projects and people, and the assumption of broader responsibilities.

CHAPTER HEADINGS

- ☐ The Transition of an Engineer to an Engineer Manager
- ☐ Engineer Manager Relationships
- ☐ The Management of Engineers
- ☐ Functions of the Engineer Manager
- ☐ Skills of the Engineer Manager
- ☐ Characteristics of the Engineer Manager
- ☐ Training for the Transition
- ☐ Effectively Performing the Engineer Manager's Role
- ☐ Summary

LEARNING
OBJECTIVES

- ☐ Learn to work with and through people to accomplish goals.
- ☐ Gain insights into his relationships to others.
- ☐ Understand the management functions required in his new environment.
- ☐ Learn skills necessary for engineering management functions.
- ☐ Understand the characteristics of the engineer manager.
- ☐ Develop ways of successfully performing as an engineer manager.

EXECUTIVE COMMENT

A. W. ANDREWS
Director, Corporate Engineering
Monsanto Company

A. W. Andrews is director, Corporate Engineering for Monsanto Company. A native of Brooklyn, New York, he holds a B.E. in Chemical Engineering from Yale University. He joined Monsanto in 1941 and since then has held numerous positions in manufacturing, research, and engineering before moving into engineering management and general management. Prior to his current assignment he was director of Oil and Gas Division (1966–72) and general manager, Plastic Division (1972–75).

The Engineer Becomes a Manager

Most of today's engineer managers at the outset of their careers did not have a conscious commitment to management as a career objective. They started in the practice of an engineering discipline and subsequently moved into management through some combination of abilities, interests, and circumstances of place and time. This being the case, the quantity and quality of preparation and training in management principles and practices which each received along the way were highly variable. Although changes in this pattern are emerging, the need for individual initiatives toward acquiring management skills is not likely to diminish.

The practicing engineer uses a combination of rigorous technical methods and personal judgment in order to accomplish the tasks at hand. The input data, the constraints, the computations, the alternatives which he tests are largely within the realm of established factual knowledge. This rigorous area occupies most of his attention. At the same time, there are very important judgmental inputs for him to make as to alternatives for study, how to provide for uncertain boundary conditions, and the like. Indeed, skills in this latter area may well distinguish the exceptional engineer from the average.

The practicing manager, however, finds the relative order of importance of the judgmental versus the rigorous technical areas reversed. While he must not lapse into technical obsolescence, his major emphasis shifts to the management skills that will determine the effectiveness of his managerial actions. Two skill areas are of great importance: (1) the ability to define the decision alternatives applicable to a given situation in a comprehensive and penetrating manner, that leads to a persuasive rationale for the actions to be taken and (2) the ability to communicate at three levels:

1. To superiors for confirmation of the action plan and commitment of resources thereto;
2. To peers to secure their active support;
3. To subordinates to secure their commitment to the chosen actions—and, very importantly, their understanding of the ultimate purposes thereof.

There is developing a body of knowledge that is relevant to the general practice of management and upon which the engineer manager can draw. It will be to the engineer manager's advantage to do so.

Today we live in a technological society. No one will deny that the engineer has been the prime mover in bringing us to this stage, but as he has done this, his responsibilities have increased. No longer is the engineer the person making a discovery in a laboratory and using it in a very local application. Instead, the engineer is having worldwide influence either by his research and development prowess or through his managerial abilities. The importance of these abilities is clearly pointed out in a book entitled *The American Challenge:* "God is clearly democratic; He distributes brainpower universally but He quite justifiably expects us to do something efficient and constructive with that priceless gift. This is what management is all about. Management is in the end the most creative of all the arts for its medium is human talent itself."[1]

Corporations of today are so technically oriented that more than half of them are headed by people with engineering educations. The bulk of the lower decision-making and supervisory jobs will soon be in their hands. This means that during the coming decade the impact of the corporation on society will literally be in the hands of engineers.

Large numbers of engineers have realized early in their careers that their conventional engineering education has only partially prepared them for the

[1]J. J. Servan-Schreiber, *The American Challenge* (New York: Atheneum 1968), p. 15.

types of positions they will be called upon to fill. Vannevar Bush expressed this well in his autobiography: "I had graduated in engineering but I was not an engineer. An engineer has to know a lot about people, the way they organize and work together or against one another. I resolved to become a real engineer. I resolved to learn about men as well as about things."[2]

The seriousness of this need for management education was shown in the National Engineers Registry Survey conducted in 1969. The summary of the data is shown in Figure 1–1.

A 1972 study by the National Science Foundation showed that in the last census over 480,000 persons with college degrees in engineering were reported under other occupational categories. For the manager, whose skill in the application of technology and knowledge of the fact that any endeavor must be managed, the importance of engineering management is beyond question. The typical engineer finds his way into an engineering department of a corporation but soon may be called upon to become a project leader or project manager, or perhaps even to leave engineering and go into the production and marketing areas of the corporation. All of these new needs call upon areas of knowledge which the engineer's college education failed to provide. What then are the problems faced by the engineer as he makes his transition into management?

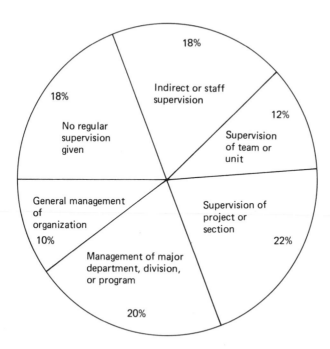

FIG. 1–1. Supervisory Responsibilities of Engineers. (Source: *Engineering Manpower Bulletin #25*, Engineers Joint Council, Sept. 1973)

[2]Vannevar Bush, *Pieces of the Action* (New York: Arno Press, 1970), p. ii.

THE TRANSITION OF AN ENGINEER TO AN ENGINEER MANAGER

The principal consideration for an engineer to enter into a supervisory role almost always is his "success and standing as an engineer." This is a very important matter because in many cases engineers demand that they be supervised by outstanding engineers. Unfortunately, this is usually the only criterion in promoting engineers to managerial positions. With such a narrow view, it is not surprising that in most cases the transition is traumatic.

One of the real problems stems from the fact that neither the engineer nor those who are responsible for guiding the engineer into management positions understand the difficulties of this transition. As a result, engineers who are promoted into management are often left to flounder in a new environment of difficulties for which they have had little or no preparation. To soften this shock and to facilitate successful transition involves training and understanding.

As an engineer manager, instead of solving his own engineering problems, he now must work as a leader of a team and concentrate efforts on getting things done through others. The engineer does not realize that if he chooses to make this change into management, the very qualities that earned him the recognition for the promotion—coming up with new and better product designs, making wise decisions in the selection of materials, seeking out improved assembly techniques, building quieter and safer products, keeping alert to new developments— are the antithesis of those required of the manager. All of these require the engineer to have the ability to work on his own and to be interested in engineering technicalities rather than in people, characteristics that are handicaps instead of assets in his role of manager.

Why is this transition so difficult for the engineer, especially the engineer who has outstanding professional engineering competence and is recognized by management as the best engineer for the promotion? Some reasons are:

1. The engineer is moving away from engineering work in which he has been most successful and satisfied. In most cases, the engineer does not even realize what is being left behind. The transition occurs at a time when the engineer has achieved an outstanding reputation in engineering and has every indication that success will continue. The engineer is now entering a new profession in which the chances of success have completely different probabilities. He was at the top of his engineering profession but now will be at the bottom of the management profession. What if he fails as an engineer manager? By the time management and he realize this unsuccessful attempt, he can no longer reenter the engineering profession at the top. Under these circumstances, the engineer manager feels insecure and fearful of making wrong decisions.

2. The engineer as an engineer manager must view problems and make decisions from a company viewpoint and not from that of a professional engineer. This transition is especially difficult for the engineer who has practiced several years. Also, this is difficult when the engineer is not prepared for this change.

As an engineer, he had certain ethical and professional responsibilities which he readily assumed. But as a manager, he must have the same dedication to helping society and mankind as to helping the company grow and be profitable. This appears to be a small change, but it is a difficult change for the engineer when adequate training and directions are lacking. The engineer must realize and appreciate that as a manager his prime function is helping achieve the company's goals and objectives and in turn providing the goods and services that society wants.

3. The engineer has a great fear that as a manager he will lose the direct control he formerly had over his work. He worked by himself and his achievements were directly the results of his own effort, but now he must work through others to achieve his goals. There are several problems for the engineer making this change:

    A. Accepting someone else's solutions to "his" problems.
    B. Dealing with people and the factors that motivate people to do a good job.
    C. Leading a team on which he does not always get his way.
    D. Learning to "trust" others.

The conditions are not very pleasing to the engineer turned manager. If they were, he probably would not have started out as an engineer. As an engineer manager, he is always at the mercy of his subordinates and their inefficiencies; therefore, he may attempt to protect himself by endeavoring to gain expertise in his subordinates' engineering speciality. Since this is impossible, in desperation he tries to become directly involved in the project. But, as a result of this effort, he is uncomfortable, gets in the way of his subordinates, and finds the day is simply not long enough to do both jobs.

4. Another change the engineer must undergo in becoming an engineer manager is to devote full time to activities that have previously ranked low in his scale of values, i.e., techniques of management. Most engineers view a typical manager as a bureaucrat and paper shuffler who is an obstacle in the way of engineers trying to do their work. They may view the engineer manager as being more interested in minor details and in building an empire than in doing good work. Unfortunately, many managers fit this description perfectly. This situation occurs when an individual is not adequately prepared and trained for management responsibilities. Rather than admit defeat and return to his old position or another one less demanding, he immediately develops work habits designed to protect his position. He becomes a bureaucrat enmeshed in paperwork, so busy that it is impossible for him to be removed from the position.

5. An engineer is trained and experienced in making decisions based on theories and laws governing the behavior of the physical world. Now in his new role, he must deal with intangibles, with rules and regulations made by people, and with the vagaries of human nature. It is an environment in which nothing is certain.

The engineer manager is much more effective in coping with the above problems if he has had proper orientation and on-the-job training. Unfortunately, there are many engineers today in supervisory positions who have not had the luxury of being prepared to take on their new responsibilities.

ENGINEER MANAGER RELATIONSHIPS   The engineer manager is a trained engineer, but immediate superiors are in many cases neither engineers nor technically trained personnel. He frequently must seek cooperation and information from other departments and groups. He finds little opportunity to control the factors affecting his actions or decisions. Many times he is not included in the total decision process and he must execute his responsibilities with less than the total available information. Certainly, the engineer manager must soon learn to think and reason in this environment, but he must be careful not to lose touch with the way of life of the engineers under his supervision. He is the link between the engineers in his department and a management that is primarily oriented toward profits. The engineer manager must at all times remain conscious of the engineers under his supervision and their particular needs.

The engineer manager must recognize the management problems in supervising engineers. It is essential that he have engineering training plus management training and be able to effectively use this training in "getting things done through others." For example, the engineer wants to have responsibility for planning his own project, even if it is a small segment of a larger project. The engineer manager must recognize and provide this opportunity. The engineer wants to be evaluated on his performance in these projects and receive feedback on his timeliness of completing projects and on the quality of his work.

The engineer manager is primarily responsible for the implementation of company policies. He must realize that to his employees he is now management. As he works through others to accomplish his goals, he is often heard to complain about his inability to get as much technical engineering work done as he would like. In fact, he is constantly looking for ways to maintain his technical engineering competence.

The new engineer manager now finds himself confronted with mountains of paperwork of questionable value. But more difficult is the unlimited number of administrative problems such as:

1. How does one get a raise for a subordinate?
2. How does one discipline?
3. Where does one requisition certain items?

These and other questions involve more than instruction. The manager needs an intelligent insight into company policies and organization. He must know and understand the reasoning behind the rules and regulations.

One difficult annual event is the performance appraisal he must do of the engineers. Management appraisal generally involves many intangible factors that the engineer may interpret as a way of avoiding the real reason for not giving a well-deserved raise or promotion. Properly done, the appraisal should include such things as how well the engineer interacted with his fellow engineers, how close his projects came to meeting deadlines, and the technical quality of his work. It is generally believed that it is best for the appraisal to be made at a time different from the time of salary adjustments. This allows the process to occur in an atmosphere in which suggestions for improvement of the engineer's performance may be better received. The engineer will accept the appraisal more readily if his performance can be measured in quantitative terms. This can be accomplished by establishing goals, the achievement of which can be measured quantitatively.

Physicians are supervised by physicians, hospital nurses are supervised by nurses, teachers are supervised by superiors with training in the teaching profession, lawyers are supervised by lawyers, but engineers are often supervised by nonengineers. The engineer often feels demoralized by this practice, causing low productivity.

THE MANAGEMENT OF ENGINEERS

Engineers are similar to other professional people, not only in how they attack and solve problems but also in how they find or seek to find a place in the organization. They respond not only to their own egocentric appreciation of the situation and their fantasies about it but also to the ways in which other engineers view the problem. Very often these are different from the ways the same circumstances are viewed by management. The ways in which management relates to the engineer guide the engineer into various working patterns that may be either productive or nonproductive.

Management preference for one organizational group over another can heighten the rivalry between engineers and management, resulting in bitter controversies. Past relationships are important to the understanding of engineers by management. Commonly experienced patterns of behavior can be considered important, for they show drives and desires that represent the individual engineer and his profession.

The ability to utilize the engineer's talents requires comprehending and gaining insight into the engineer's way of reacting and relating to problems. This requires an objectivity about the engineer's feelings and behavior that comes only with experience. It is among the most difficult skills required of management.

Unfortunately, relationships—particularly new relationships—contain elements of earlier ones; evaluating new acquaintances is based on experiences with others. Management often sizes up new engineers by consciously or unconsciously fitting them into the patterns of others they have known before. An astute engineer manager with knowledge and understanding of the individual engineer may recognize subtle similarities to others. The intense relationship between engineers becomes an integral part of their personality and it influences the entire behavioral pattern. Distortions of their relationship with management are corrected by an engineer manager as he gains maturity and a secure ego identity. The work patterns established by the leadership of the engineer manager should strongly direct an individual engineer toward an understanding of his role and function. Otherwise, there is a fundamental tendency throughout the course of the work to pour the new engineer into an old mold; he is dealt with according to past experiences. Such repetitive tendencies are usually more prevalent among managers without engineering experience.

FUNCTIONS OF THE ENGINEER MANAGER

Engineering management functions are those tasks which are performed by the engineer manager. These include (1) preparing budgets, (2) training subordinates, (3) developing policies, (4) evaluating projects, and (5) supervising subordinates. Some of these functions were performed when the engineer manager was working as an engineer in addition to the tasks specific to his engineering specialty. For example, the engineer may have been required to do budgeting, reporting, policymaking, and program assessment for his projects. Being an engineer manager, however, involves not only performing engineering management functions, but also major changes in the following:

1. The *frequency* of performing various engineering management functions.

2. The *enjoyment* or *dislike* of the engineering management functions. Various engineering management tasks might have been distasteful when he did them as an engineer, but they become enjoyable when he does them as an engineer manager (or vice versa).

3. The *difficulty* of accomplishing various engineering management functions. The functions might become either more or less difficult when performed as an engineer manager rather than as an engineer.

4. The *time* required, for engineering management functions now demand more time for the engineer manager than when they were performed as an engineer.

5. The *importance* given engineering management functions when they are performed as an engineer manager rather than as an engineer.

6. The *character* of engineering management functions when one performs them as an engineer manager rather than as an engineer.

Some of these changes create real problems for the engineer during the transition into engineering management. Therefore, it is necessary to examine what the changes really involve and what really occurs when the engineer becomes an engineer manager. In terms of functions, the engineer manager is required to perform more extensive managerial work than when he was an engineer. Many times, the engineer does not enjoy performing these functions because they interrupt his normal work activities. If, however, the engineer manager can be helped to realize and thoroughly understand the importance of these activities, he generally finds they give satisfaction. The importance of these functions is given in Table 1–1.

TABLE 1-1

Importance of Engineering Management Functions

Engineering Management Functions	Reasons for Importance
Preparing plans and budgets, developing policies	The engineer manager must develop goals and objectives, and methods of achieving them, obtain resources and facilities, and provide leadership.
Evaluation of project	This provides the engineer manager a means of assuring that goals and objectives are being achieved and that his department is making a contribution.
Training and supervision	This enables the engineer manager to assist subordinates to be productive and to develop an efficient staff, thus assuring the achievement of the goals and objectives of the organization.

Certain engineering management functions were particularly difficult for the engineer manager when he was an engineer. These management functions and the reasons for their difficulty are given in Table 1–2. The cause of these difficulties may be the lack of necessary aptitude or training or because the function seems completely unrelated to the engineer's specialization. If the engineer manager does not understand the reasons for and importance of performing these engineering management functions, he often sees them as preempting his time for performing more worthwhile activities.

TABLE 1-2

Difficult Engineering Management Functions

Engineering Management Functions	Reasons for Particular Difficulty
Preparing budgets and reports	No aptitude for the function, lack of training, unrelated to engineering, causes anxiety and conflicts, very time-consuming
Preparing plans, developing policies, making evaluations	Lack of quantitative information, many unknowns requiring subjective judgments that cause the engineer manager a great degree of uncertainty
Training and supervision	Involves subjective judgments (engineers prefer objective situations), many interpersonal relationships but no training in how to handle and cope with them

SKILLS OF THE
ENGINEER MANAGER   To perform effectively, the engineer manager must develop the
engineering management "skills" needed to actually perform
the engineering management functions. These skills are referred to as *engineering management skills* to distinguish them
from engineering skills.

Engineering management skills are useful in performing many, if not all, of
the engineering management functions. They are, for example, the ability to
communicate ideas, to develop organizations, to work with people, and to solve
problems. The new engineer manager must realize the relative importance of
using these skills in successfully performing his work. Other engineering management skills that are important include:

1. Working within the organizational structure
2. Coordinating department goals and activities
3. Leadership
4. Decision making
5. Creativity
6. Developing confidence as an engineer manager

As with the engineering management functions, there are two major
changes for the engineer with respect to engineering management skills:

1. More of the skills are applied with greater frequency as the engineer enters into engineering management.

2. The skills take on a broader scope in the engineer manager role than in that
of the engineer.

For example, the engineer who deals with financial activities does not go beyond
specific projects, but the engineer manager is financially responsible for projects
encompassing a whole area or group, sometimes involving negotiations with
other groups in the organization. These greater responsibilities carry the need for
the engineer manager to be involved with a larger number of individuals at higher
management levels of responsibility and with a wider range of occupations
and backgrounds.

These skills are recognized as important to those actually in engineering
management; however, top management generally does not realize the difficulty
the engineer manager has in acquiring them. Middle and top management have
these skills and experience little difficulty in using them.

The most definite change in the role of the engineer manager is the shift
from task-centered skills to skills required to work within the organizational

structure. This requires the engineer manager to become more skilled in understanding the organization itself—its goals, objectives, procedures, and policies. He must also develop skills in working with managers who have substantially different backgrounds and interests from those of engineers. The important difference between engineering and engineering management is that engineering skills are centered on technical problems and engineering management skills are people-centered. These differences are compared in Table 1–3.

TABLE 1–3

Skills Required of Engineers Compared to Skills Required of Engineer Managers

Engineers	Engineer Managers
Solving technical problems	Working within the organizational structure
Application of techniques	Fiscal analysis
Technical communication	Dealing with personnel
	Working with people other than engineers
	Leadership
	Coordination
	Communications

As the engineer manager assumes greater responsibility, the above engineering management skills increase in importance and become sources of difficulty. The supervisors, however, generally tend to underestimate the difficulty of acquiring these important people-centered skills.

The engineering management functions to be performed and the skills needed for effectively performing them have a great potential for creating problems when an engineer becomes an engineer manager. If these problems are recognized, less difficulty is experienced. An engineer generally accepts the view that engineering management functions are necessary activities of the organization; this is especially true of those functions he performed as an engineer.

CHARACTERISTICS OF THE ENGINEER MANAGER

The new engineer manager may have difficulty in recognizing the completely different patterns of behavior associated with being an "engineer manager" compared with being an "engineer." The key differences between the roles of the two positions should shed light on the situation. The differences in roles are given in Table 1–4.

TABLE 1–4

**Differences in Roles between Engineers and
Engineer Managers**

Engineer's Roles	Engineer Manager's Roles
Originates projects	Evaluates projects
Creates, seeks new ideas	Provides facilities to help engineers
Works on specific programs	Does overall planning
Has limited responsibilities	Has responsibility for a department or group
Is specialized, is technically oriented	of people
Obtains facts himself, is objective	Is people-oriented, is responsible for and
Utilizes own skills	responsive to people
Has limited concern for finances	Motivates others
	Utilizes skills of others to achieve goals
	Has fiscal responsibilities

The characteristics of an engineer center around his technical specialty and his interest in using technical knowledge and skills. An engineer manager, however, has much broader interests involving both operations and the exercise of authority. The engineer in transition needs to understand these different characteristics, which are given in Table 1–5.

TABLE 1–5

Characteristics of Engineers and Engineer Managers

Engineer's Characteristics	Engineer Manager's Characteristics
Liking new and different things	Being a leader
Directly attacking problems	Making decisions
Being independent	Wanting to give assistance and guidance
Being recognized	to others
Exercising technical knowledge	Helping achieve the firm's goals and
and skills	objectives
Building relationships with other	Gaining authority over and support from
engineers	others

The changing roles from engineer to engineer manager create a major change in rewards expected by each role. These rewards are given in Table 1–6. Both groups of rewards contain achievement satisfactions and status satisfactions for the engineer and engineer manager. The rewards for engineers, are mainly technically oriented; the rewards for engineer managers are primarily those of broader activities with greater authority and responsibility.

TABLE 1-6

Rewards for Engineers and Engineer Managers

Engineer's Rewards	*Engineer Manager's Rewards*
Successful job performance	Position in the organization
Freedom of action	Achieving major goals and objectives
Financial rewards	Financial rewards
Challenge of a difficult task	Supervising the work of others
Satisfaction of creating	Involvement in major decision making
Professional relationship with colleagues	Promotion to higher levels of responsibility and authority
Engineering achievement	Assuming responsibilities with requisite
Social status	authority

The attitudes toward the rewards associated with engineering management that influence the engineer as he considers transition are many and complex. Among the favorable factors are:

1. Opportunity to develop and acquire new expertise, knowledge, and experience
2. Financial advancement
3. Prestige of position
4. Opportunity to accept a wider scope of work that provides a change of pace and emphasis
5. Opportunity to direct projects and achieve engineering goals and objectives

Some of the unfavorable factors are:

1. Risk of moving from an established, successful environment
2. Uncertainty about the aptitudes or abilities required
3. Uncertainty of the degree of satisfaction in the new job
4. Lack of opportunity to perform engineering work

Thus, the favorable factors of engineering management can be summarized as providing the engineer broader opportunities in influence, challenge, and personal advancement. The unfavorable factors causing the engineer to be reluctant to enter engineering management appear to be leaving one's engineering specialty and a feeling of uncertainty about being able to perform the engineering management role effectively.

The favorable and unfavorable attitudes toward the rewards of the engineering management position produce stress and tension for most engineers. If the engineer is long established in his position, the rewards may have little appeal;

in fact, they may have a negative influence on his desire to assume management responsibilities.

With respect to the transition to management, three general groups of engineers are recognized:

Group 1. Engineers who have essentially "engineer manager's characteristics," although they are working at the moment in their engineering specialties. The "engineer's characteristics" have no deep involvement for them, though the pattern must be operative to some degree. One could expect these engineers to be relatively active in their efforts to move into engineering management.

Group 2. Engineers very involved with engineer's characteristics and somewhat reluctant to move into engineering management. Once the transition has been made, however, they discover that engineer manager's characteristics can have meaning for them; satisfactions are derived which had not been anticipated. Most are reluctant to make the change, but after working as engineer managers, they "begin to love it."

Group 3. Engineers for whom engineer manager's characteristics have a definite negative appeal. If these engineers should go into engineering management (either for financial advancement or because of being "drafted" into it), they would most likely find the experience frustrating and their performance would suffer.

Many engineers look favorably on the prospect of going into engineering management while they are still working as engineers. One can imagine that a majority of engineers in engineering management are in Group 1 and that relatively small numbers are in Group 2 and Group 3. Management can, through a self-selection process, reduce the number of Group 3 individuals going into management. Candidates who lack the characteristics associated with engineering management tend to avoid the opportunities to take these positions. Unfortunately, every organization occasionally promotes a Group 3 engineer to a senior supervisory position as a reward for excellence in research and then finds it must "protect" him from engineering management tasks by assigning him a managerially oriented deputy or assistant. When this occurs, too much emphasis has been given to his technical qualifications or competence for the promotion instead of to his engineering management abilities such as leadership, motivation, organizational planning, decision making, problem solving, communications, and achievement of organizational goals.

TRAINING FOR THE TRANSITION Beginning with the era of "scientific management," it was recognized that the role of supervision was distinct enough to call for some type of training or educational effort.[3] Usually, it is the modern offspring of such early training programs as are offered to engineers in their first managerial or supervisory positions. Training programs have been developed largely on the basis of management intuition and personal experience—particularly the experience of se-

[3]Fredrick W. Taylor, *Scientific Management* (New York: Harper & Row Pub., 1947).

nior officials who have a continuing interest in the problems of transition. Management has recognized for some time that the transition is difficult and chancy and that organizations should provide some assistance to individuals going through the process. Most attention is devoted to providing some type of training for new engineer managers.

The *key training need* falls in the area of the personal skills required to perform the engineering management role. The specific problem areas are sharply defined. The new engineer manager needs to learn how to cope with or manipulate the organizational environment, to coordinate group effort, and to lead others. There is also some need for training in specific aspects of the engineering management functions. This is especially true for budgeting, financial analysis, and project evaluation. Too frequently, knowledge about the organization and its principal personalities is assumed, and therefore it is not covered in supervisory training programs. There is a need for better understanding of the informal structure and processes of the organization.

What kind of training would be best suited to fill these needs? A program based on intellectualized and cognitive approaches, such as lectures and readings, is not sufficient. In formal training programs there is often a tendency to use case studies and laboratory materials which are not easily related to the work environment of the organization. This should be avoided.

The experiences needed are those which tend to expose the individual directly to the organization. These experiences can be provided by an internship program. Another advantage of such pre-entry or internship activities is that they let the engineer discover the extent to which he can obtain motivational satisfactions from the managerial role.

The understanding that a new engineer manager needs for a successful transition comes from a combination of experience, observation, and assistance from his superior. This last item appears to be a critically weak link. Therefore, if management is serious about helping engineers in their transition to engineer managers, they must:

1. More strongly emphasize the responsibility of supervisors for developing their subordinates
2. Provide incentives for greater concern on the part of supervisors in meeting these development needs
3. Provide the means by which supervisors can effectively exercise such concern

EFFECTIVELY PERFORMING THE ENGINEER MANAGER'S ROLE

The engineer manager must understand and appreciate the rewards that follow from overcoming the problems of transition. The rewards are many, including the reward of seeing people grow and develop professional capability under the engineer manager's guidance and leadership. Another is the feeling of accomplishment in seeing a complex project completed under the engineer manager's direction which no one engineer could complete on his own.

There is the satisfaction of directing and planning the technological advancements to achieve maximum success. Another reward is learning that working with people can be as much fun as developing a highly technical engineering product.

When the engineer moves into an engineering management position, he probably receives a raise in pay. In most cases, however, the engineer believes he is still being paid to do engineering with only a dash of supervision thrown in. His company may also consider him just a high-priced engineer. But soon he is involved with budgets, meetings, performance reviews, and reports; and he finds that little time is left to perform the technical functions in which he had established himself. This is a critical period in the transition experience.

Because engineers are trained to concentrate on objectives and the means to achieve them, they can effectively utilize their analytical skills in planning, organizing, and controlling a complex unit of the company. The immediate supervisor should be in the best position to give the new engineer manager the necessary counsel. Unfortunately, the supervisor may be little wiser in many of these problem areas than the new engineer manager. His concern and interest will go far toward eliminating a number of problems new engineer managers have as well as stimulating the firm to provide the necessary preparation for new engineers about to make the crucial move into engineering management.

Engineers should understand that their passion for exact answers and unequivocal solutions sets them apart from other competent people in nonengineering occupations. Most of the management people the engineer manager will come in contact with will not use specific, scientific logic. He is always tempted to try to mold others to conform to his own pattern of thinking and reasoning. The result could, however, be a breakdown of cooperation between engineer manager and nonengineering management. Therefore, the engineer manager, without surrendering his professional competence or personal integrity, must learn to let others achieve goals and objectives in their own way rather than imposing the engineering way.

SUMMARY   A principal qualification used in considering an engineer for a supervisory role is often the engineer's "success and standing as an engineer." While this technical ability is frequently an important factor when determining who should fill supervisory or managerial positions, neither the engineer nor those involved in guiding them into management realize the difficulties of this transition. To soften this shock for the engineer and to facilitate successful transition involves training, understanding, and drastic attitude changes.

Engineering management functions are the tasks performed by the engineer manager. These include preparing budgets, training subordinates, developing policies, evaluating projects, and supervising subordinates. Many of these functions are not unfamiliar because they were performed when he was working as an engineer, in addition to the tasks of his engineering specialty.

Skills in engineering management are necessary for an engineer manager to perform effectively. These skills are referred to as engineering management skills to distinguish them from engineering skills. As with the functions of engineering management, there are changes that take place: (1) more of the skills are applied with greater frequency as the engineer enters into engineering management and (2) the skills take on a broader scope in the engineer manager's role than in the engineer's role.

The engineering management functions to be performed and the skills needed for effective performance may create problems for the engineer making the transition to engineer manager. Less difficulty is experienced if these problems are recognized. An engineer will generally accept the view that engineering management functions are necessary activities of the organization if he relates them to those functions he performed as an engineer.

Management has recognized for some time that the transition is difficult and chancy and that organizations should provide some assistance to individuals going through the process. Key training is needed in the area of personal skills required to perform the engineering management role. The new engineer manager needs to learn how to cope with or manipulate the organizational environment, to coordinate group effort, and to lead others.

The engineer manager may look forward to the rewards that follow from overcoming the problems of transition. These rewards are many, including seeing people grow and develop professional capability under his guidance and leadership, a feeling of accomplishment in seeing complex projects completed under his direction which could not have been done by a single engineer, and satisfaction in directing and planning major technological advancements in order to achieve maximum success.

Important Terms

Engineer Manager:
One who holds managerial responsibilities for technical activities related to achieving corporate objectives.

Objectivity:
A characteristic that the engineer manager exhibits in observing technical problems with a company viewpoint instead of strictly an engineering viewpoint.

Supervision:
The activities of the engineer manager to translate the plans and strategies into action through his subordinates.

Transition:
The period in the engineer's professional career during which he changes from performing engineering activities to performing engineering management functions.

For Discussion

1. What should the company do to help the new engineer manager adjust to his position?
2. What are some of the major differences between the job performed by engineers and the job performed by engineer managers?
3. What factors make it difficult for an engineer to make the transition to an engineer manager?
4. Compare the skills of an engineer and an engineer manager.
5. Describe the differences in attitude toward a project by an engineer and an engineer manager.
6. Describe the attitudes or traits of an engineer that make the engineer unqualified to be a manager.
7. What are some of the reasons the transition from engineer to engineer manager is so difficult?
8. Does a good engineer necessarily make a good engineer manager?
9. What are the rewards of being an engineer manager?
10. Why do we need engineer managers in today's society?
11. How should an engineer manager relate to fellow engineers?
12. Make an attempt to clarify the difference between the requirements for an exceptional engineer and the requirements for an exceptional engineer manager.
13. The key training need falls in the area of the personal skills required to perform the engineering management role. Define the problem areas and discuss how each can be mastered.

CASE 1-1

The New Engineer Manager

Tom Johnson, Vice President of the Metal Products Company, a manufacturer of highly technical products, was contemplating the increase in sales volume during the past year and the sales forecast. It was now apparent that the firm needed to create a position of Director of Research and Development (R&D). If this position were approved, the individual would be responsible for the day-to-day operation of the R&D Division and the long-range planning for new product areas.

Tom thought of offering the job to Rex Smith. Rex had been with the company for 15 years and had worked his way up from a junior engineer to chief engineer. Although he has not completed his bachelor's degree, he is practically a genius when it comes to the engineering and technical problems of manufacturing for Metal Products. Tom was told that Rex would be very disappointed if he were not given a shot at the job and that it would probably have a severe effect on his morale. Tom was sympathetic to Rex's feelings. After all, Rex had been extremely loyal to the company and had done a good job over the years.

Nevertheless, Tom had some real doubts about Rex's ability to handle a true engineering management job. Tom knew that being a top-notch engineer did not guarantee that Rex would be a good engineer manager. Tom had heard complaints from several of the firm's engineers about their feelings that Rex was narrow-minded and that he tried to handle people the same way that he handled machines.

1. Should Rex be promoted?
2. What would be some of the consequences of promoting Rex?
3. What would the consequences be if Rex were not promoted?
4. If Tom promotes Rex, what should Tom do to try to ensure Rex's success as an engineer manager?

CASE 1–2

Promoted Engineer

After Rex Smith had completed his first year as a supervisor of the engineering department of Metal Products Company, he sat down to review his progress. This was his first true managerial job although he had led small groups of engineers from time to time during the preceding 10 years. As he reviewed matters, he felt that he had performed fairly well. He did have some question about whether or not he had maintained the morale of his people. He felt that he might have suffered from the fact that he had been one of the group for a number of years and in moving up to this position he had bypassed Jim, who was somewhat older and better educated and felt he should have had the job. During his first year as supervisor they had been civil to one another but he felt that Jim's productivity was declining. He had been somewhat reluctant to discuss the problem with him.

Rex also had another problem: John was nearing retirement age. Although John was a highly competent individual, his productivity was certainly not what it had been in past years. Rex was a little concerned about how to handle this situation. He felt awkward about it because John was so much older than he. Rex felt comfortable with technical problems but uncomfortable with people problems. He knew that he had tended to push this problem under the rug hoping it would go away.

As he looked back over the year, however, he felt he was probably going to have to do something about both of these problems before they had an effect on other younger engineers in his department. He already had some complaints about distribution of workload.

1. How should Rex deal with the problem of the engineer who felt he was being bypassed?
2. How should he handle the problem of the engineer nearing retirement?
3. How can he protect the morale of his younger engineers under these circumstances?

For Further Reading

ALDEN, JOHN, "Earning More and Enjoying It Less?" *New Engineer* (March 1977), 31–34.

BADAWAY, M. K., "Prescription for Success," *Industrial Research* (June 1976), 100–103.

CHAMOT, DENNIS, "Professional Employees Turn to Unions," *Harvard Business Review,* 54, No. 3 (1976), 119–27.

DALTON, GENE, et al., "An EE for All Seasons," *IEEE Spectrum* (December 1976), 42–46.

FULLER, DON, "Motivation Is More Than Gimmicks," *Machine Design* (November 11, 1976), 116–20.

HAAVIND, ROBERT C., and RICHARD L. TURMAIL, *The Successful Engineer-Manager.* New York: Hayden Book Company, 1971.

HEIMER, ROGER C., *Management for Engineers.* New York: McGraw-Hill Book Company, 1958.

IMBERMAN, WOODRUFF, "As the Engineer Sees His Problem," *The Conference Board Record* (April 1976), 30–34.

"Managing Engineers Is a Challenge in Leadership," *Iron Age* (May 10, 1976), 54–56.

RAUDSEPP, EUGENE, "Teamwork: Silent Partner in the Design Group," *Machine Design* (August 7, 1976), 62–64.

——, "When and How to Delegate," *Machine Design* (January 8, 1976), 66–69.

RUBEY, HARRY, et al., *The Engineer and Professional Management.* Ames, Iowa: The Iowa State University Press, 1970.

STEINMETZ, LAWRENCE L., "The Over-Motivated Engineer," *Machine Design* (October 7, 1976), 98–101.

TURNER, BARRY T., *Management Training for Engineers.* London: Business Books Limited, 1969.

THE TRANSITION TO DECISION-MAKING RESPONSIBILITIES

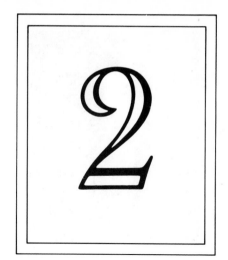

Making decisions is not a new experience for the engineer, but the engineer manager must augment technical information with financial and personnel information in making decisions. This requires him to develop additional decision-making techniques.

CHAPTER HEADINGS
- ☐ Engineer's Decisions Assume New Dimensions
- ☐ Decision-Making Process
- ☐ Techniques of Choosing Between Alternatives
- ☐ Timing of Decisions
- ☐ Types of Decisions
- ☐ Obtaining Acceptance of Decisions
- ☐ Principles of Decision Making
- ☐ Summary

LEARNING OBJECTIVES
- ☐ Recognize new dimensions of the engineer manager's decisions.
- ☐ Learn the steps of the decision-making process.
- ☐ Gain insights into techniques to choose between alternatives.
- ☐ Know how to gain acceptance of decisions.

EXECUTIVE COMMENT

RICHARD J. REDPATH
Division Vice President and Director
Corporate Engineering
Ralston-Purina Company

Richard J. Redpath joined Automation Engineering as a project designer in 1961. He was manager, Automated Systems, Singer Company and national director of engineering, Johnson and Johnson before joining Ralston-Purina in 1977. In his present position of division vice president and director, Corporate Engineering, he is also responsible for the management of corporate engineering and resource conservation.

The Engineer's Transition to Decision-Making Responsibilities at the Management Level

Often during planning sessions in which promotional candidates are discussed, an individual will be judged as not being management material. Generally, the individual in question has desires in the management career area. The peer group clearly feels this individual is unskilled in the decision-making process required for that management position. Examples of how something was handled in the past are offered in evidence. The person's career is subsequently directed toward technical decision areas.

The individual so judged generally is bewildered and later asks, "What can I do to be eligible?" Unfortunately, scientific method engineering training cannot cover all situations. The engineer is trained and disciplined to gather the facts, to measure them, to make comparisons, and then simply pick the best alternative. This works very well with inanimate objects and systems but not always with ill-defined subjective data.

A good engineer manager is an effective decision maker. The successful manager very often has little direct control over the specifics of a project. The manager may not even have a choice in what is to be done or who is to carry out the tasks. This seems contrary to his being able to make good decisions, and it is at the root of our technical man's problem. The technical man assumes he must make the choice. He spends a great deal of his time gathering facts and attempting to understand much of the detailed data relevant to his project. A complex project involves many different areas of engineering, e.g., civil, mechanical, electrical, and financial. The project may have unusual constraints and legal considerations. The client may view the end product differently. The local plant manager may desire changes to the plan requested by the general manager of the division. Many combinations of decisions are possible. Compromises in one area may provide greater opportunities in another.

In spite of what seems to be an impossible task, some engineers do very well at managing complex programs. Many top executives in operating companies have engineering degrees, and the trend appears to be growing. Conversely, the establishment of technical ladder, nonmanagement priority positions has occurred in many large companies. The modern company is attempting to identify people with potential management skills early. The more sophisticated companies are exposing their high-potential candidates to early career experiences requiring more than $F = ma$ technology decisions. The technically well-trained engineer finds himself trying to understand a nontechnical person's needs. The management trainee encounters several options to decide from, and there is no clear best solution. Some errors are made but the person who takes charge and applies his best judgment generally averages positive.

The business environment allows continual retesting of decisions. The successful engineer manager is one who does not irreversibly lock in a decision. The manager must not assume that he is the only source of data for the decision. The good manager will solicit many opinions, have empathy for his client's needs, and, most important, be pragmatic about what is possible. Sound decisions are arrived at by listening to all of the input, not just the hard facts. Once the decision is made, careful communication must follow to complete the loop with all those involved.

Decision making at the management level is a learned process. It starts when the new engineer first experiences the consequences of decisions being made by others above him. It becomes evident to him that many things are assumed rather than proven. The experienced manager has developed certain skills through many past similar situations. Through his experience, he can add judgment factors to the fragmented data and opinions. The engineering manager must learn to insert various human beings into the decision criteria.

The transition to an effective managerial decision maker cannot be completely taught through a graduate engineering management program. Communications and interpersonal skills are important, but perhaps most critical is a sense of team functioning. The person who puts aside his ego and accepts another's input, who keeps an upbeat attitude toward the objectives, and who enjoys stimulating others in productive activity has made the transition to a good management decision maker.

The process of deciding what to do is fundamental to an engineer. In fact, most of the engineer's activities involve decisions of varying importance; therefore, decision making is natural to him. An engineer's entire education and training are oriented toward solving problems and making decisions and also toward performing the actions necessary for implementation. An engineer may be viewed as a professional knight in the art of deciding. The decisions the engineer must make are critical because mistakes are expensive.

The engineer is familiar with the process of solving problems and making decisions. He is well acquainted with the scientific method and uses it in his decision-making process. This leads him to follow a very precise step-by-step process for solving problems and making decisions. Thus, the engineer views the decision process as a rational ordering, from the setting of objectives to the determination of the final course of action. It provides a logical process from the beginning to the end. Generally, the engineer's role is to provide a single decision as part of a larger decision-making process.

Now as an engineer manager he must be concerned about the entire problem, which includes being responsible for the outcome of the decisions. But now decisions may be altered by the actions of others, perhaps requiring him to alter the logical sequence of his decision-making pattern. Without realizing it, he may soon be practicing a nonsystematic decision-making process instead of the systematic process he used as an engineer.

ENGINEERS' DECISIONS
ASSUME NEW
DIMENSIONS

As an engineer manager, one of his primary functions will be to make decisions that determine the future direction of the department and of other parts of the organization over both the short and long term. These decisions will involve various activities in the organization, including production, engineering, finance, marketing, and personnel. To complicate the decision-making process, most decisions involve several of these functions. Thus, he has now added new dimensions to his responsibilities compared to those he had as an engineer. However, these new dimensions can be either very rewarding and satisfying to the engineer manager or a most frustrating experience. When frustration occurs, he may soon acquire the attitude that everything he does will go wrong and be a "bad" decision. As a result, he avoids making decisions. His subordinates will find it impossible to get a decision from him. He will take the position that time will take care of everything. It will, but he will no longer be in control, for the decision will be in the hands of someone else.

Any engineer manager can become a better decision maker by following a systematic approach to solving the problems confronting him. Poor decisions are made when he considers the decision too unimportant to give it adequate attention or avoids using a systematic approach in reaching a decision. A systematic decision process allows him to make rational decisions. It attempts to establish a logical framework based on the scientific method, which results in the selection of a course of action from among various alternative actions.

In the decision-making process, the engineer manager must always consider the factors and functions that are under his control. In fact, he will feel he has little control over anything, but actually he will have control over the majority of his activities. The successful engineer manager should always bring into his decisions all of the factors over which he has control.

Before a decision can be made, an alternative must exist. The process of decision making selects from alternatives the course of action to be carried out. If there are no alternatives, there can be no decision. The alternatives and the criteria for comparison are necessary functions of the decision-making process. This seems very straightforward, but, unfortunately, most decisions must be made in an environment of uncertainty. Each alternative has both desirable and undesirable aspects that must be reconciled. Since the engineer manager is unsure of future results, how can he obtain these? How sure is the engineer manager about obtaining the results of each alternative or even identifying each alternative correctly? All of these situations present interesting and challenging opportunities for the engineer manager.

For some areas, the level of knowledge required and the complexity of the decisions to be made are clear and straightforward and the criteria are well defined. Information is obtainable, future outcomes are predictable, and risks are understood. In these situations, decisions seem scientific, mathematical, and almost automatic. The engineer manager is very comfortable and generally makes successful decisions in this environment.

In many situations, however, the engineer manager finds vague criteria, which, of course, makes comparability difficult. The prediction of risk and future performance may be difficult. As a result, judgment must be used to balance this vague information, assess risk, and select a course of action—a process for which he is not trained. He may feel insecure and try to avoid the decision-making process. The engineer manager faced with making this type of decision must attempt to create an appropriate environment. Achieving this environment requires a clear and straightforward definition of the decision-making factors. It requires the engineer manager to spend greater time and effort in the decision-making process so that vague situations are better defined and proper criteria are developed.

The importance of following and building on established practices and experience is well illustrated in the following story. During an oral examination for the Ph.D. degree a young doctoral candidate was asked by one of the more venerable of the assembled scholars the following unlikely question: "Assume that the desk here is a stove and that wastebasket over there is a bucket of water. Now, I ask you, how would you solve the problem of heating the water on the stove?"

The degree candidate was stunned for a moment and then replied. "Why, I guess I would pick up the bucket of water from the floor and place it on the stove."

To the candidate's surprise, the old professor was visibly pleased. "Excellent, excellent!" he exclaimed. "Good bit of reasoning!"

"Now that you have solved that problem so neatly, let me ask you a more difficult one." With this the professor walked to the wastebasket, picked it up, and placed it on the table on the far side of the room.

"Suppose the bucket of water were over there. You agree, do you not, that this is a more difficult problem?"

The puzzled degree candidate managed a nod of agreement.

"Very well then, how would you solve the problem of heating the water now?"

After a moment's hesitation, "Well, I guess I would pick up the bucket of water from over there," said the candidate, making a gesture in the direction of the table, "and then place it on the stove."

To this the old professor stormed, "Wrong, wrong, wrong! And you think you're a scientist, do you? Ha! Well, let me tell you, damned little progress would be made if all scientists went about their work the same way you went about this problem! Reduce the difficult problem to a simpler one that you've already solved. That's the secret to progress." Then with a slight smile he added, "Why, any scientist worth his salt would have known enough to move the bucket from the table to the floor first. Putting the bucket on the floor, you see, reduces the problem to one you have already solved!"[1]

At this point, the degree candidate was at a loss for words. However, the old professor had made an excellent point for the engineer manager making a decision: reduce a complex problem to several simpler ones that he already knows how to solve. More often than not we attempt to look at the whole problem as a new experience instead of solving it from this point of view.

DECISION-MAKING PROCESS   For the engineer manager, the following steps are suggested for reaching sound decisions:

1. *Recognize the problem.* One of the first steps in the decision-making process is for the engineer manager to recognize that a problem exists. This appears to be a relatively simple step, but to recognize and realize that there is a problem is very important. To do this requires the engineer manager to be constantly aware of all operations

[1]C. Springer, R. Herliky, and R. Beggs, Basic Mathematics (Homewood, Ill.: Richard D. Irwin, 1965), pp. 135–36.

and activities, and to compare these with plans, past experiences, standards, and other operations, so as to know that a problem exists. Many times problems exist but the management of the organization does not recognize them.

A change of some kind either in the enterprise itself or in the external environment usually is the causal factor behind the decision problem situation. An adjustment may be necessary because something occurs that alters the existing state of affairs and requires attention. Or managers may take systematic steps to bring about planned changes of existing operations to avoid stagnation. Planned obsolescence by automobile manufacturers is an example of this type of decision making. Or a decision may be brought about by an undesirable deviation from expected performance, as evidenced by changes in quality or equipment operation. Problems of the latter type crop us constantly in all functional areas and at all levels of management in an organization. One no sooner solves one problem than another presents itself. Frequently a solution of one problem creates a new or potential problem. Therefore, managers are continually involved in making decisions and solving problems. Some problems are pressing and require immediate attention. Others can be handled at a later date. Some are complex and broad in scope; others are simpler and more limited.

The important thing is for the engineer manager to operate in such a manner that he is alert to problems as they arise so that appropriate measures can be taken to correct them. These measures will involve making decisions.

2. *Define the problem.* The next step is for the manager to define the problem in terms he and his subordinates can understand. For example, the manager recognizes that the project is behind schedule. This certainly is a problem, but it must be defined. When this is done, several problems may be in evidence. The project is behind schedule, but what are the reasons? These must be defined in terms that have meaning and are useful to the employees of the organization. There may have been equipment breakdowns on the project, materials may not have been ordered with sufficient lead time, or the wrong materials may have been made available for the project. Each of these problems must be analyzed to determine the real cause of the delay. It may be that labor is involved, which can be further defined in terms of poor morale, poor supervision, or union–management relations. In this step, the objective is to define the problem in as specific terms as possible so that everyone can comprehend it.

The philosopher John Dewey said that a question well put is half answered. Thad B. Green and others have echoed the view that the problem definition is the key to effective problem solving.[2] A manager has to know what he is trying to do and where he is trying to go before he can begin to accomplish his objectives. A problem resembles an iceberg in that what is visible actually represents only a small portion of the problem. It is important to define the problem clearly enough that one gets to the root of it and does not just treat the symptoms.

3. *Obtain relevant data.* The engineer is well trained in securing relevant data. Now as an engineer manager this experience becomes very helpful and is an important step in the decision-making process. This involves collecting quantitative

[2]Thad B. Green, "Problem Definition—Key to Effective Problem Solving," *Management Advisor* (November–December 1973), 42–45.

data from actual experiments, accounting records, project records, statistical reports, government reports, and other publications. The accounting department is an important source of data; however, it maintains records primarily for the Internal Revenue Service and not always in a form useful for decision making. Nevertheless, the accounting department can maintain records to meet the engineer manager's requirements if it is told what data are needed and what format is required.

In addition, nonquantitative data must be gathered through interviews, questionnaires, and reports. This is difficult for the engineer manager to evaluate since he has been trained to use quantitative information. This difficulty can be overcome by rating the individual data on a scale of, say, from 1 to 10 as shown in Figure 2-1. The nonquantitative data are placed in a quantitative format and can be included in decision-making models. This technique can be used in many cases in which one may think only value judgments can be used. For example, in design decisions, the characteristics of each design being considered can be rated on such a scale so that the relative values can be included in the decision process.

4. *Evaluate and analyze data.* During the past several decades, many new methods have been developed which engineer managers have extensively used to analyze data. Most of these techniques have been developed in the area of operations research using computer technology. Many of these techniques provide the engineer manager with the information to optimize decisions which provide the best answer to complex problems. To the engineer manager, many problems will seem so complex and involved that they are capable only of suboptimization, i.e., being optimized as a part of a process, not the whole. Although these techniques provide answers, they are not necessarily the best answers because of the complexity of the problem(s) and the great variance of some factors. In these situations, the engineer manager must use his experience and judgment to arrive at a decision. He can sometimes assist the decision-making process if he breaks the problem into small segments appropriate for analysis.

In analyzing the data, the manager must look to the past when similar problems might have occurred. If the equipment worked under similar conditions before, why will it not work now? What conditions have changed that have caused the problem? Frequently, problems can be solved quickly if suitable comparisons with the past are made.

5. *Formulate alternatives.* This step involves formulating alternative courses of action (Figure 2-2). The engineer manager must formulate as many feasible solutions as possible to ensure that sufficient courses of action are considered. If his first course of action later becomes inappropriate, he has other alternatives avail-

FIG. 2-1. A Rating Scale for Evaluating Data Through a Range of Important to Unimportant

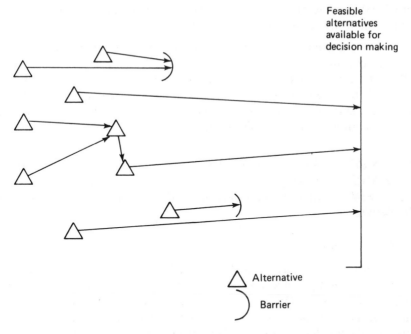

FIG. 2-2. Choosing Feasible Alternatives for Decision Making

able which he can immediately implement. This is necessary because he is managing in a dynamic environment in which conditions are constantly changing and influencing the situation. Therefore, an alternative that is the best possible choice today may not even be valid tomorrow. Having other alternatives available allows the engineer manager to avoid making decisions in a crisis situation when adequate time or resources are not available. Frequently, the decision maker finds limiting factors standing in the way of accomplishing his desired objectives. When searching for alternatives, he may want to confine his search to those which overcome the limiting factors. This may lead him to choose alternatives which will be most helpful in achieving his goal. Limiting factors are not always easy to find, but their importance in developing appropriate alternatives makes the effort worthwhile.

6. *Select a course of action.* Using the preceding steps, the engineer manager is in a position to make a decision on which course of action to follow. Often he will still find himself with insufficient data yet forced to make a decision and be responsible for the outcome. Unfortunately, he does not have perfect knowledge of the future, yet he is still required to make decisions. Deciding which course of action to follow is difficult, but uncertain situations make using the decision-making process more important than ever. At this stage of the decision process, one helpful technique is an effective cost and profit analysis. This analysis will provide the engineer manager with information on which course of action will result in the lowest cost or highest profits.

7. *Review the decision.* As already pointed out, the engineer manager is operating in a constantly changing environment. Therefore, the chances that the course of action he has selected will work out as planned are rather unlikely. It is then necessary to review the decisions periodically to ensure that the desired results are being achieved. If not, there will be new problems that will require his immediate attention and new decisions that must be made to achieve his objectives.

TECHNIQUES OF CHOOSING BETWEEN ALTERNATIVES   Since choosing alternatives is at the heart of the decision-making process, let us look further at some of the techniques that have been developed to handle the more complex problems.

Decision Trees

One of the best ways to analyze the decision is to use a decision tree. A decision tree makes it possible to see the directions that actions might take from various decision points. Obviously, adequate information is seldom available to make an accurate decision at any given time. The tree depicts future decision points and possible chance events. One can also improve the usefulness of the tree by adding probabilities that various uncertain events will happen.

A common problem facing the engineer manager from time to time is new product introduction. There is always a substantial risk that a new product will not be successful. Decisions have to be made on whether to invest significant sums of money in a plant to produce the product or to hedge one's bets by modifying some existing equipment to produce the product on a temporary basis. The higher the investment made in new equipment in which a more efficient process will be available, the lower will be the costs. Conversely, the "jerry-rigged" process may have relatively high costs. The manager can use the decision tree to analyze the probable results of his decisions.

An example of such a tree is shown in Figure 2–3. It will be noted that permanent tooling for the product would cost $500,000 and profits are estimated to increase $200,000 a year over the life of this product under the best of conditions. Two other alternatives are shown: (1) $75,000 profit per year under less ideal conditions and (2) $100,000 loss in the event that the product would fail. If temporary tooling is used, the increase in profits is less because the quantity of product available would be reduced and is thus set at $50,000 per year for the ideal conditions. If somewhat less sales result, then the gain of $15,000 per year is estimated; if the product fails, a loss of $25,000 per year is indicated.

As the engineer manager surveys this array, his next concern is the probability of each of these three events happening. We have assigned probabilities of 0.6., 0.2, and 0.2 to the two options with three events respectively. The last column indicates a multiplication of the probability times the various estimated pro-

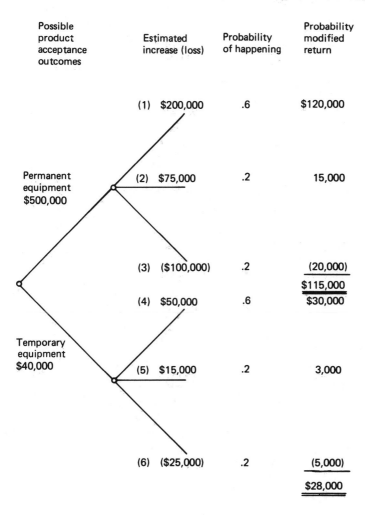

Possible product acceptance outcomes	Estimated increase (loss)	Probability of happening	Probability modified return
	(1) $200,000	.6	$120,000
Permanent equipment $500,000	(2) $75,000	.2	15,000
	(3) ($100,000)	.2	(20,000)
			$115,000
	(4) $50,000	.6	$30,000
Temporary equipment $40,000	(5) $15,000	.2	3,000
	(6) ($25,000)	.2	(5,000)
			$28,000

FIG. 2-3. A Decision Tree for an Equipment Investment Project

fits. It can be seen in this case that even under ideal conditions, the return on the investment is 24% with permanent tooling but with the temporary tooling it is 75%. Since a failure of the product under permanent tooling would provide a significant capital loss, it might well be that the manager would choose to go initially with the temporary tooling. There is another reason why he might do this. Generally a new product requires several years to reach its maximum sales. By starting initially with the temporary tooling, it is possible to develop the market and get a better handle on what capacity would be needed for the long run. With this information available, plus what has been learned by debugging the equipment, the manager can make a better decision on what equipment to purchase for the new plant and how large to build it.

The temporary tooling may also allow the opportunity to test out the possible competitive pressures that will exist. Who becomes a competitor might well influence the future course of the company. Koontz and his colleagues have made an interesting analysis of a more complex version of such a tree under conditions of competition.[3]

Preference Theory

Modern decision theory has provided some techniques using statistical probabilities that bring into play individual preferences. Given the data on the decision tree, there is still the question of the decision that the manager will make. With a 70% chance of success, one would normally think that he might be led to go in that direction. But there is a 30% chance that he might be wrong; and since his future in the corporation and his family's livelihood are at stake, he might not want to take such a risk.

The theorists have tried to develop the probabilities of different types of decision makers' accepting risks. They feel that the size of the risk is a factor, as well as the level that the manager occupies in the organization. It is generally felt that higher-level managers are more accustomed to taking larger risks than lower-level managers are, and that by the very nature of their decisions, they tend to involve larger elements of risk. It is also possible that the type of decision being made will influence the amount of risk that is willing to be taken. Managers may be more willing to gamble on such things as advertising than they are on plant investments.

It seems important, then, that the engineer manager analyze very carefully the decision to be made and the chances of its success. He must recognize that if he does not take risks, he will miss opportunities. It is probably reasonable to expect that in the course of his working life he will be somewhat more successful if he is willing to make choices when the chances for success are only in the 60% to 70% range.

Operations Research

To approach some of the more complex problems, techniques have been developed that utilize mathematical models in computers. We will briefly review some of the techniques involved. In-depth information on the procedures can be found in some of the references given at the end of the chapter. Some of the major decision-making techniques include:

1. *Mathematical techniques.* Mathematics provides methods of evaluating the many alternatives for accomplishing objectives in the most efficient possible manner. This involves combining resources and activities that will optimize the

---

[3]H. Koontz, C. O'Donnell, and H. Weihrich, *Management*, 7th ed. (New York: McGraw-Hill, 1980), pp. 265–66.

organization's goals. Note that there must be alternatives for these techniques to be used. One allocation technique is transportation programming for resources over an area. Another more familiar technique is linear programming, which is an effective tool when the objective function is to maximize profit or minimize cost and can be defined as a linear function with limited resources. In most cases, linear relationships do not exist in the real world, but the analysis involves only a small portion of the curved production function which can be assumed to be linear. If no linear condition exists, nonlinear programming techniques are required. When problems have a sequence of interrelated decisions and the engineer manager needs to evaluate these over an extended time period or number of events, dynamic programming is a particularly appropriate approach to these types of problems.

2. *Network techniques.* Networks are effective in the analysis of projects which determine priorities in activities that require special attention and for monitoring the activities or the project undertaken. Gantt charts, PERT, and CPM are techniques available to the engineer manager. These techniques will be discussed in Chapter 10. Flow process charts, operation charts, activity charts, man–machine charts, X-charts, and R-charts are similar techniques for use in the decision-making process.

3. *Inventory control techniques.* The inventory problem requires decisions to balance shortage costs and ordering costs, thereby answering the question of how much to order at one time and when or how many times to place an order.

4. *Waiting-line techniques.* These are techniques for evaluating the optimal number of service facilities for customers arriving at some random distribution so that the cost of service and cost of waiting are balanced.

5. *Simulation techniques.* Simulation provides the engineer manager with information on alternative courses of action with respect to decision relationships. Through simulation, the engineer manager can introduce constants and variables associated with the problem, establish possible solutions, and use criteria to measure their effectiveness. Many variables and constants associated with the problem are similar to a real-world environment.

6. *Markovian techniques.* Markov process models are useful in studying the evolution of certain systems over repeated trials, and hence the probability of the system being in a particular state at a given time period.

7. *Heuristic program techniques.* Heuristic techniques may be used on poorly defined problems when specific relationships cannot be developed. These techniques are appropriate for complex operating problems and they provide information to determine the effect of a decision.

8. *Game theory.* This technique deals with situations in which conflicts are present. An attempt is made to determine the behavior rules that will provide the most favorable outcomes based on the value of alternative outcomes that the decision maker may determine for each of these. The engineer manager will encounter many such instances in which these readily occur, such as collective bargaining, pricing, products, and competition.

TIMING OF DECISIONS   The engineer manager must realize that there is a proper time to make decisions and implement them. When he delays or postpones decisions, it causes serious problems or may even jeopardize the success of the entire organization. A delay in ordering parts for equipment or supplies, for instance, may require a production line to be halted. There may be a number of reasons why the engineer manager may delay making an important decision and these may all be valid. But on occasion he tries to avoid making a decision because he hopes that the problem will soon go away. This delay can be damaging both to him and his group. But, on the other hand, quick improperly developed decisions generally turn out to be poor decisions. It is important, therefore, for the engineer manager to plan his decision making so that he will have adequate time and not be hurried.

The manager will never have enough data but must always be prepared to make decisions on the available information. Sometimes he may be able to delay for a short time in order to obtain important additional information. He must be certain, however, that in taking this course of action he is not merely avoiding making the decision. After a reasonable period of time, he must use his best judgment to arrive at a decision.

Proper timing of decisions requires the engineer manager to be constantly alert to events. For example, equipment technology continues to improve. If the engineer manager purchases a machine just before new technology is introduced, the machine will be obsolete before it arrives at the plant. If new technology occurs or new models are introduced every five to six years, then purchasing new equipment late in the cycle can be an unwise decision.

A reversal of the situation occurs when the engineer manager is involved in introducing equipment. He may not want to be the first on the market with new technology, choosing to let others pioneer the introduction with its attendant problems and costs. Or he may want to enter the market after customers have been identified and product characteristics have been established.

The engineer manager must attempt to foresee problems which may become serious at a future date. Not all such problems will be of equal importance. Therefore, he must be able to evaluate the possibilities and select those problems which are likely to affect the organization in the future.

The engineer manager is required to make decisions in many different areas. It is important that he understand which decisions should be delegated to the various organizational levels. This is an important consideration because it provides a way to train engineers for further promotion to managerial levels.

TYPES OF DECISIONS

For the routine day-to-day decisions, standard procedures can be used. Usually decisions of a routine nature are made at lower levels of the organization because uncertainty and risk are less. Decisions that are routine and defined by standard procedures involve procurement of standard items, production control, cost control, and quality control.

Some decisions occur infrequently and involve a new set of alternatives each time. They are difficult to analyze and are not susceptible to standardized procedures. They require the engineer manager to use judgment, experience, intuition, and even rules of thumb. Such decisions are involved in the introduction of automated equipment, plant layout, research and development for new products and processes, and purchasing computer equipment.

The variables involved in decisions that occur infrequently are usually more complex than those of routine decisions. Decisions relative to research and development and new products and processes are complex because they involve economic, sociological, psychological, and ethical factors. All of these are difficult to treat in the quantitative terms with which the engineer manager is familiar. Routine decisions are much easier to make because data for them are generally available in quantitative terms, and the factors affecting the decisions are easier to control. The engineer manager sees all of the factors affecting these infrequent decisions as uncontrollable. Data are difficult to obtain and results are difficult to forecast. The value system affecting the alternatives is subject to opinions, objectives are conflicting, and probabilities of occurrences are difficult to determine.

OBTAINING ACCEPTANCE OF DECISIONS

Overcoming resistance to decisions and learning how to make changes smoothly depends on the group's degree of involvement in the decision process. Subordinates frequently resist changes because their status in the company may change or their positions or even their livelihood may be threatened. The engineer manager is obliged to make decisions on old and new product lines, changes in equipment, and changes in employees' operating conditions. These decisions, as well as almost all others, require employees to make changes. Subordinates may resist decisions that require change and at times do everything to make them unworkable.

The task of obtaining acceptance of decisions, therefore, is just as essential as making the decisions. Unless the engineer manager can convince his subordinates to accept his decisions, all his efforts are in vain. Obtaining acceptance of decisions is equally important in furthering his own self-confidence. Each time he is successful, he finds it easier to gain acceptance for the next decision. If he is unsuccessful in obtaining acceptance, then most of his decisions will not be implemented even though they are sound. His ideas may be brushed aside by his subordinates and he becomes an ineffective leader.

The engineer manager must take the steps necessary to ensure that his decisions are accepted. He may do this in several ways, including the following:

1. By determining the response of those involved during the decision-making process. This involves learning the goals or objectives of the people directly involved, since one's personal goals have a major influence on how he responds to company decisions. Most people work in behalf of their personal goals. One way to learn these goals is for the engineer manager to listen instead of talk. However, understanding the individual background, cultural heritage, and social needs of his subordinates is helpful.

2. By requesting the subordinates involved to act as instructed. He may use either a reward or a punishment, depending on whether or not instructions are followed.

3. By knowing the subordinates' goals and objectives. He can thus determine the degree of acceptance of his decision and probable reaction.

4. By being an effective leader through keeping in touch with what is going on in the organization, making a series of good decisions, and maintaining high standards of integrity.

5. By getting subordinates involved as part of the decision-making process so that they will support the decision wholeheartedly. When the engineer manager brings to his group a proposed solution, it should be presented as a tentative one so that all have a chance to think it through. If it is proposed as the final solution, the reaction will most likely be negative. This will certainly be the result when the proposed solution appears to endanger the subordinates' personal goals. The discussion will then lead to all the unwanted consequences that will be developed in implementing the decision and to none of the benefits.

If the employees are allowed to think about and discuss the proposed decision, acceptance is more certain. However, the engineer manager must keep an open mind to questions and comments. Often in an exchange he may find that his initially proposed solution is really unworkable. If he has stated it as a tentative one, he can back off without losing prestige with the group and, in fact, may gain respect because of his flexibility and willingness to consider their suggestions.

With this high degree of participation, each employee is likely to feel that the decision is his own, for each has had a part in developing it. Making contributions and participating in the overall solution are great motivators. Through this participation technique, the engineer manager has strengthened the respect and loyalty of his group.

6. By obtaining acceptance through effective communication. This involves informing people of the decision, keeping them informed throughout implementation, and maintaining feedback. The engineer manager must communicate the decision in terms that all can understand. One thing the engineer manager must realize is that this phase is hard work and will never be completely done. He can communicate effectively by using the following steps:

    A. Determine what information is required by the individuals.
    B. Use the best sources of the information.
    C. Inform the individuals involved.
    D. Ensure that there is feedback.

These steps provide the engineer manager with the means to convey useful information to his subordinates. The more effective the communications made by the engineer manager, the more likely people will accept his decisions. When an individual is informed directly and immediately, it makes the individual feel important and an integral part of the organizational goals.

Even the most successful engineer manager experiences trouble in making decisions. Decision making is hard work and at times it must be realized that no decision is the best one. However, he is required to make decisions; this is one of his duties. He should keep the following principles in mind:

PRINCIPLES OF DECISION MAKING

1. The engineer manager must *recognize* and *define* a problem before a decision can be made.

2. A decision should help to *achieve* or *contribute* to a specific dynamic goal.

3. The decision-making process has an *order* that should be followed by the engineer manager.

4. The basic procedure for decision making is the *scientific method.*

5. The engineer manager must always *develop alternatives* for use in the decision-making process.

6. The engineer manager must get as many as possible *inputs from subordinates and superiors.*

7. The decision-making process must be *flexible* and *responsive* to an ever-changing environment.

8. There is *more than one* satisfactory decision for most problems.

9. Decision making requires *time* and *effort.*

10. A decision requires that *action* must take place.

11. Generally, *poor decisions* are made during crises.

12. The engineer manager needs to view problems as *opportunities* for the organization.

13. The world is *dynamic* and *constantly changing;* therefore, the engineer manager must constantly reevaluate past decisions.

SUMMARY The engineer manager finds that his decisions are more complex and broad in nature than they were when he was an engineer. To approach them appropriately, he finds it necessary to use a formalized technique based on the scientific method. It involves recognizing the problem, defining the problem, obtaining relevant data, evaluating and analyzing the data, formulating alternatives, selecting a course of action, and reviewing the decision.

Of these, the most difficult one is evaluating alternatives. Decision trees offer a way of assisting in an evaluation under conditions of uncertainty. A number of new techniques have been developed through operations research to handle the more complex problems. These involve mathematical, network, inventory control, waiting-line, and simulation techniques.

It is important that decisions be made at the appropriate time. They may be made either too early or too late and the manager must be very careful not to become a victim of either extreme. One of the more difficult problems the manager has is selling his decisions to his subordinates. It is useful if he arranges to have involved them somewhere in the decision-making process because it will make the decision more acceptable to them.

Important Terms

Alternatives:	Feasible courses of action that are available for the solution to the problem.
Decision:	Selection of an alternative course of action from among two or more that are available.
Decision-making process:	An orderly process used for arriving at the most desirable solution to a problem.
Forecast:	Determination of the events that will occur in the future.
Systems approach:	Concern with the total problem as it affects manufacturing, marketing, and engineering.

For Discussion

1. Discuss the benefits and dangers of pioneering the introduction of a new product.
2. Discuss the factors involved in the decision to replace or repair and modify a piece of production equipment.
3. What are the important steps in the decision-making process?
4. What techniques (models) are available to the engineer manager to aid in decision making?
5. What can the engineer manager do to assure acceptance of his decisions?
6. How can the anecdote about the doctoral candidate be used to direct the engineer manager?
7. When an engineer is promoted to be an engineer manager, will there be any changes in the nature of his decision making? What are they? How should he make his decisions in his new position?
8. What is the importance of the timing of decisions for an engineer manager?
9. Describe three of the operations research decision-making techniques available to the engineer manager.
10. Is it true that most of the time there is "only one feasible way of accomplishing a course of action"? Discuss the importance of alternatives.
11. Are there any significant drawbacks in using the scientific method for decision making? List the advantages and disadvantages.
12. Develop a decision tree for the case of a company operating at full capacity and with product demand rising. The choice is between new equipment and overtime. A 20% rise in sales is anticipated, bringing an increase in profits of $300,000 with new equipment and $150,000 with overtime. But volume might drop 5%; therefore, there are two possible results. If there were a sales drop, profit would be $220,000 with new equipment and $175,000 with overtime. The probabilities are 60% for a sales increase and 40% for a sales decrease.

Case 2–1

The Decision

Rex Smith must decide whether to purchase or lease a new piece of laboratory test equipment. He calls Mary Todd to his office and says, "Mary, we are in need of a mass spectrograph for the analysis of some new products resulting from our research work. This is an expensive piece of equipment and I would like you to prepare an analysis and present your recommendation on whether we should lease or buy. I'll need an answer in two weeks."

1. What information will Mary need for her decision?
2. Outline an appropriate approach to arrive at a decision.

Case 2-2

Financial Difficulties

Tom Brown, one of Rex Smith's engineers, visits him one morning with a tale of financial woe. Tom's story seems very real and he asks for a raise. Although Tom does satisfactory work, he certainly doesn't deserve a raise by comparison with others in the department. Since the engineers are not constrained to a specific salary range, it is possible to give him a raise.

1. What action do you recommend that Rex take?
2. What will the consequences be?

Case 2-3

Too Much

Rex Smith has another problem in his department. One of his older engineers (55 years old) with 20 years of service with the company has developed a severe drinking problem. He is usually late for work and usually has a hangover. His productivity is low and he frequently takes a half-day off.

After discussing the problem with him, Rex arranged for him to receive professional help. It proved to be to no avail. They felt there was little chance for improvement.

The man is competent when sober but his actions do have an effect on the department. Rex ponders what to do?

1. Should Rex fire him?
2. What other actions might he take?

For Further Reading

ALEXIS, M., and C. Z. WILSON, *Organizational Decision Making.* Englewood Cliffs, N.J.: Prentice-Hall, Inc., 1967.

BIERMAN, HAROLD, JR., et al., *Quantitative Analysis for Business Decisions,* 4th ed. Homewood, Ill.: Richard D. Irwin, Inc., 1973.

BROSS, IRWIN D. J., *Designs for Decisions.* New York: The Free Press, Macmillan, 1966.

BROWN, REX V., et al., *Decision Analysis for the Manager.* New York: Holt, Rinehart & Winston, 1974.

EILON, SAMUEL, "What Is a Decision?" *Management Science* (December 1969), B172–89.

HALL, JAY, et al., "The Decision Making Grid," *California Management Review,* 7, No. 2 (1964), 43–54.

HARRISON, E. FRANK, *The Managerial Decision Making Process.* Boston: Houghton Mifflin Company, 1975.

HILLIER, FREDERICK S., and GERALD LIEBERMAN, *Operations Research,* 2nd ed. San Francisco: Holden-Day, Inc., 1974.

HYMAN, RAY, and BARRY ANDERSON, "Solving Problems," *Science and Technology* (September 1965), 36–41.

JONES, MANLEY H., *Executive Decision Making.* Homewood, Ill.: Richard D. Irwin, Inc., 1962.

KARGER, DELMAR W., and ROBERT G. MURDICK, *Managing Engineering and Research.* New York: Industrial Press, Inc., 1969.

KEPNER, C. H., and B. B. TREGOE, *The Rational Manager.* New York: McGraw-Hill Book Company, 1965, Chaps. 3–9.

MACARMMAN, KENNETH, "Managerial Decision Making," in *Contemporary Management,* ed. Joseph McGuire. Englewood Cliffs, N.J.: Prentice-Hall, Inc., 1974.

MCCREARY, EDWARD A., "How to Grow a Decision Tree," *THINK* (March–April 1967), 13.

PATZ, ALAN L., and ALAN J. ROWE, *Management Control and Decision Systems.* New York: John Wiley & Sons, Inc., 1977.

POUNDS, WILLIAM, "The Process of Problem Finding," *Industrial Management Review* (Fall 1969), 1–19.

RADFORD, K. J., *Managerial Decision Making.* Reston, Va.: Reston Publishing Company, Inc., 1975.

REYNOLDS, WILLIAM, "Problem Solving and the Creative Process," *MSU Business Topics* (Autumn 1967), 7–15.

SANFORD, EDWARD, and HARVEY ADELMAN, *Management Decisions: A Behavioral Approach.* Cambridge, Mass.: Winthrop Publishers, Inc., 1977.

TERSINE, RICHARD J., "Organization Decision Theory: A Synthesis," *Managerial Planning* (July–August 1972), 18–24.

PERFORMING THE MANAGERIAL FUNCTION

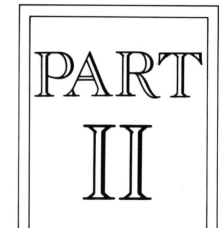

PART II

THE TRANSITION TO PLANNING TECHNICAL ACTIVITIES

The engineer has been taught to plan his engineering work, but now as an engineer manager he has overall responsibility and his plans must now provide the basis for the actions of others. It is essential that they complement the company's goals or objectives.

CHAPTER HEADINGS
- ☐ The Nature of Planning
- ☐ The Planning Process
- ☐ Techniques of Forecasting
- ☐ Implementing Plans
- ☐ Plans in Operation
- ☐ The Do's of Planning
- ☐ Summary

LEARNING OBJECTIVES
- ☐ Gain insight into the importance of planning.
- ☐ Learn to develop a plan.
- ☐ Gain knowledge of various techniques of forecasting.
- ☐ Understand methods for implementing plans, including development of policies and procedures.

EXECUTIVE COMMENT

LLOYD E. REUSS
Vice President
General Motors Corporation
General Manager
Buick Division

Lloyd E. Reuss joined General Motors engineering staff in 1959. After holding a number of positions in Chevrolet ranging from development engineer, through chief engineer of Camaro, chief engineer of Vega, and divisional manager of product planning, in 1975 he was promoted to chief engineer for Buick. He held this Buick position until 1978 when he was promoted to director of engineering for Chevrolet. In late 1980 he returned to the Buick Division as vice president and general manager.

Planning by Engineer Managers

Planning is an essential part of all engineering activities. At the project level the planning is operational: the plotting of a course of action in advance to achieve a defined end product. A sequence of necessary events that must take place ties the plan together, usually with a series of bench marks established. In effect, the individual engineer's plan is almost a schedule and is highly structured.

When the engineer moves into management, planning takes on a different meaning, a meaning that becomes broader as the engineer manager's responsibilities increase. Here, planning is a method of guiding managers so that their decisions and actions affect the future of the organization in a consistent and rational manner and in ways aimed at meeting organizational objectives. For the engineer manager, planning is not a means of minimizing risks but rather a means of maximizing opportunities at an acceptable risk.

In his role as a manager, the engineer becomes responsible for directing the activities of others and, more important, responsible for their interactions as a team and guiding that team to success. The engineer manager must plan in such a manner as to provide direction, thrust, and motivation to his people. The people within the team must be allowed to use their creative talents.

One pitfall of the upward moving engineer in transition to management is to continue to use the successful day-to-day highly structured planning technique and apply it to his new responsibilities. The team then usually becomes just an extension of the new manager with him enmeshed in large amounts of detail. In all likelihood, much creative talent of his subordinates will not be used.

Rather, his management plan should emphasize the strategic rather than the operational. Broad but clear objectives must be established for the longer term. In turn, these goals must be communicated to the groups involved with assurances that the intent is understood. Communication is fundamental to strategic planning along with knowing essential facts, establishing realistic time frames, and being able to monitor progress.

In doing this, the engineer manager must stress looking at where the organization is going and draw on the innovation of his people for the detailed execution of the various programs aimed at accomplishment of the broader objectives.

In today's changing, dynamic business world, it is no longer possible to justify a lack of planning and a delay in decision making on the basis of "as long as we are getting somewhere, that's what counts," or "if we do not do anything, the problem will go away." It is getting somewhere—where you want to go and with specific routes and timetables—that counts. For this reason, the engineer must become committed to formal planning.

THE NATURE OF PLANNING

A crucial first step for every engineer manager is to clearly understand the elements of planning. Planning is deciding in advance what to do, how to do it, when to do it, and who is to do it. It is the thinking process that should be carried on before action is taken. Planning is an important function of supervision and one in which skills can be learned and specific principles applied.

The planning function is normally performed by the engineer without the realization of what is actually taking place. His formal training requires extensive study, thought, and planning to perform design functions—steps very similar to those required in the engineering management planning function. Coming up with new and better product designs, making use of new materials to meet certain design specifications, seeking out improved assembly techniques, designing quieter and safer products to meet government standards, and incorporating new developments are all assumed to be part of any design function and within the engineer's normal responsibility. He must determine the requirements of the system and develop specifications and plans for how the system or product will be developed.

From these initial broad specifications, the engineer begins to design elements of the system. For each element, specifications become more detailed, but still within the broad outline. As the design progresses, he must be concerned with how the system will be manufactured in the plant. At times designs and specifications must be modified to meet various production requirements. After the system has been designed and constructed within these specifications, it must be tested in the laboratory and perhaps further tested in the field under actual operating conditions. As a result of these tests, the system may undergo redesign and additional changes.

Even after the system is in production, the engineer is constantly redesigning the system to improve its efficiency and reduce its cost. This effort causes the engineer to constantly look for new techniques. To the engineer, the gaining and application of new ideas are essential to fulfilling his function in the organization. He recognizes that the design process is a very critical and important phase of engineering. But, when he becomes an engineer manager, the planning function seems like a completely new experience. He does not immediately realize that it is similar to the engineering design process. As an engineer manager, he deals with similar activities in the planning function but with a much broader perspective than as an engineer.

The engineer manager establishes objectives in a way similar to that when he was an engineer establishing specifications for a system. As objectives are achieved, they may be modified, just as when he modified technical specifications.

The engineer comes alive when he sees the actual production of his design, as does the engineer manager when he sees his plans implemented to achieve established objectives. Even as the engineer is constantly evaluating and redesigning based on the results of laboratory experiments, so the engineer manager is constantly adjusting plans to achieve established objectives resulting from internal and external changes in the operating environment (Figure 3–1). This requires anticipating changing conditions through the use of forecasts and having alternative plans prepared for use when these changes occur. The engineer manager is constantly developing better plans or changing or abandoning plans that will not achieve his objectives, just as the engineer is constantly developing better products through incorporating new ideas and techniques. If the engineer manager is able to see the similarity between technical planning and the planning required in his new managerial position, the new role becomes an exciting and rewarding experience.

This chapter will present technical planning concepts useful for the engineer manager in relating his engineering experience and knowledge to the requirements of engineering management planning. Because of their training and experience as engineers, most engineer managers do an excellent job of technical planning, once they realize what is involved.

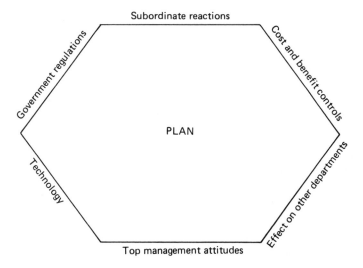

Subordinate reactions

Government regulations

Cost and benefit controls

Technology

PLAN

Effect on other departments

Top management attitudes

FIG. 3-1. Plans Are Influenced by Many Forces

THE PLANNING PROCESS

The engineer manager may view the planning process simply as a rational approach to the future (Figure 3–2). If the future were a certainty, planning would be relatively easy. However, many factors (both internal and external) are constantly changing, making it difficult, if not impossible, to forecast the future with complete accuracy. Because of this, alternative plans must be developed to be put into practice when original plans cannot be achieved. This is the same process he performed as a practicing engineer when experimental results dictated changes in the design of the system.

The steps in the planning process are similar to the steps in the decision-making process outlined in Chapter 2. This is because decision making is a major part of planning. But the environment is somewhat different and thus it is worthwhile to review the steps from a planning perspective.

Step 1: Recognizing Opportunities

The starting point for planning is the awareness of the need to take action or to seize an opportunity. In a sense, planning is a type of crystal gazing into the future. Its beginning involves developing broad statements about what the engineer manager wants to achieve, or it might be defined as developing a purpose or mission. To do this requires an ability not only to visualize future opportunities but more importantly to see them clearly and completely in order to realistically evaluate the strengths and weaknesses of each situation and the expected outcome.

A business may have the social purpose of producing and distributing economic goods and services. It may accomplish this by fulfilling the mission of producing certain lines of products. It may have decided on these lines by looking for

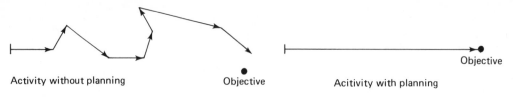

FIG. 3–2.   Unplanned Versus Planned Activities

product opportunities consistent with its raw materials and technical know-how. In this manner it has seized an opportunity. Any organization must have a mission before its component parts can set objectives.

Step 2: Setting Objectives

Objectives are the ends toward which activity is aimed. Asking some of the following questions will help to provide the basic information needed to develop appropriate objectives:

1. What action is required to achieve the mission? Specific description of the facilities and equipment to perform the proposed activities must be provided.

2. Why? Only required activities are included and all unnecessary activities are eliminated to ensure the best arrangement of activities.

3. Where? A designation of the areas in the department where each activity will be accomplished must be made.

4. When? People and facilities must be available at the proper time. A definite beginning and ending time must be established. This involves identifying each separate activity and the overall plan.

5. Who? Establishing assigned duties and responsibilities requires the determination of the particular skills and abilities available. If particular skills or equipment are unavailable, can they be obtained?

6. How? The manner for achieving the objective must be established through reviewing the above questions. This serves as an overall evaluation of the entire plan for completeness and direction toward achieving the desired objectives.

After the engineer manager has determined the enterprise objectives by following the above process, he can be confident that they are feasible objectives at this stage of the planning process. These objectives will then provide guidance in developing the objectives of specific departments or groups.

It is important to recognize that objectives can only be achieved through appropriate activities which are directly under the control of the engineer manager. He must be careful, however, not to confuse objectives with activities.

Step 3: Developing Information

Forecasts of the expected environment in which the engineer manager will operate are made at the time of setting objectives. Forecasting is the process employed to determine the future. Alternative plans are prepared so that regardless of what occurs, objectives may be met. For the engineer manager, this not only involves using other groups' forecasts but also developing his own. The marketing department prepares forecasts of sales, prices, and product needs; the engineer manager is concerned with forecasting technical development, production costs, new equipment, and labor needs.

Since forecasting the future environment is complex, it is unrealistic to attempt forecasting every detail of the future. For this reason, forecasts are limited to elements that will have the most influence on operations. Certain assumptions must also be made; for example, that a major war will not occur during the next ten years or that inflation will continue at a certain rate.

It is impossible to achieve complete agreement in forecasting the future, whether at the national level or in a corporation. Since agreement to use a common forecast is important to successful planning, it is a major responsibility of the engineer manager to make sure that his subordinates understand the forecast with which they are expected to plan. It is not unusual for individuals in a well-managed organization to have different views but through group deliberations arrive at a forecast that they can all accept.

Step 4: Developing and Evaluating Alternative Plans

The engineer manager will need alternative plans, since we can seldom forecast the future correctly. Generally, he will not have problems developing alternatives, but determining the most feasible alternatives may be difficult. It is necessary for him to reduce the number of alternatives to those that promise to be the most productive.

The engineer manager wants to develop plans that will be successful. To make realistic decisions about specific plans, he must be selective and have an early warning system that will advise him of changes in the financial, technical, or organizational environment. This requires predicting conditions within the operating environment, both internally and externally, and foreseeing which parts will cause problems. He wants to develop plans in which the unit's assets and shortcomings offer the most advantages with the least handicaps. The ideal plan would be one that has minimum areas where problems may occur but still enables the organization to operate successfully.

An appraisal of the strengths and weaknesses of the alternative plans helps the engineer manager to visualize the consequences of following any one of them. Such consequences are brought to light by studying the effect they might have on the people, production, financial, and management resources. One plan may appear to be the most profitable but involve high risk and a large investment. Another may be less profitable but have less risk. Both plans must be viewed in the light of whether or not they complement the firm's long-range objectives.

In appraising alternative plans, the engineer manager must compare each plan in terms of the tasks that must be accomplished for each phase of the plan. He then must appraise the group of people responsible for each of these tasks to see if they are able to accomplish their assignments. Next, the engineer manager should determine whether the ability to perform tasks has improved over the past few years, has stayed about the same, or has fallen to a lower level. Performance trends point out areas of group performance that are likely to continue strong or to need attention if the plan is implemented. He may determine that the weaknesses can only be overcome at great cost and therefore the plan is not economically feasible.

Many times plans are selected in a hit-or-miss fashion; they may be the result of a pet idea or of following what someone else suggested. The potential for success or failure begins to emerge when the plans are compared. A viable plan with appropriate alternatives can be developed only through examination in the light of forecasts and the available resources of management, people, facilities, and finance.

Step 5: Selecting the Plan

The last step involves selecting the plan. The engineer manager must now make a decision from among alternative plans which may appear equally acceptable. His decision involves considering both tangible and intangible factors. The intangible factors, i.e., size or length of commitment, degree of flexibility, degree of certainty, and human aspects, are factors that the engineer manager must consider. After these are reviewed, a cost analysis of major features is made in which the contributions of each feature are compared to their effectiveness in serving planned objectives. From here the cost of each valid alternative is developed. Reliance on past experiences probably plays a larger part than it should, especially with the experienced engineer manager. Past successes and failures, however, make an excellent guide to future decisions. Certainly experience is very valuable and must be carefully analyzed along with other information.

The final plan selected will in due course be budgeted. If the plan is for the overall enterprise, income statements and balance sheets will be prepared. Capital programs will be developed. If for an engineering department, its expense budget will be developed as part of the whole.

TECHNIQUES OF FORECASTING Forecasting the future environment must be recognized by the engineer manager as important in the planning process. Forecasts are much more than formal hunches or guesses about the future. Good forecasts are based on analyzing the behavior of selected factors related to the economy, technological changes, social and political demands, and business conditions as indicated by sales.

It is necessary for any business enterprise to develop an overall view of the future by forecasting. Engineer managers at various levels in the corporation will be involved in such a forecast. Here we will review several types of forecasts that would normally be synthesized into an overall corporate forecast. As we discuss each of them, we will briefly review the associated analytical techniques.

In predicting the future, it is necessary to study both the past and the present. Any forecast involves an analysis of past trends, extrapolation of these trends into the future, and the adjustment, refinement, or revision of the projected trends to allow for any factors expected to cause deviation from the established trends. Such analyses have a therapeutic effect on the organization as a whole. Reviewing the past and studying the present compels managers to look carefully at the decisions they have made in the light of changing environmental conditions. It may well disclose areas where necessary control is lacking. It forces the entire organization into a more unified manner of thinking. Then when they focus their attention on predicting the future from the past, it helps them to develop a singleness of purpose which is important to planning.

Even though much emphasis is placed on forecasting, it must be recognized that all forecasting is subject to a degree of error, for the best analyses or judgments cannot result in clairvoyance. There is a certain amount of guesswork in all forecasting. Managers must recognize the unavoidable margin of error in such prophecies and not expect too much from them. Nevertheless, it is important, that they examine the underlying assumptions of the forecast and determine if they are supported by facts or reasonable estimates and if they are in tune with corporate policies and plans.

Economic Forecasting

Economic forecasts are based on employment, productivity, national income, and gross national product. Information on these areas is regularly available from the President's Council of Economic Advisors, the Wharton School, The Conference Board, and many university and bank economists. There is no shortage of national and regional economic forecasts. The major problem is knowing which ones to use.

These economic forecasts are developed by calculating the gross national product (GNP) based on forecasts of population, productivity increases, unemployment percentages, and average workweek. Allowances must be made for components of the gross national product made up of government purchases, personal consumption expenditures, business fixed and inventory investment, residential construction, and other investments.

The corporate economist, after studying these available national and regional economic forecasts, can translate them into their impact on the industry which his corporation serves. Further analyses can then develop the impact on the company itself.

Despite the fact that the national and regional forecasts are broad in nature, businesses have found that they can relate reasonably well to these projected movements. Obviously, some companies or industries will relate better than others. For example, the home construction industry correlates well when it adjusts housing starts for the lagtime between starts and sales.

Input-output tables have been found useful by some companies in developing forecasts of their markets.[1] These tables show the relationships of industries to one another and their sharing of gross national product by calculating the purchases and sales made between industries. The steel industry was one of the first to make use of this type of analysis. Due to the lack of government funds, there has been some interruption of data available since the concept was initiated

[1]M. F. Elliott-Jones, *Input-Output Analysis: A Non-Technical Description* (New York: The Conference Board, 1971).

in 1947. There has been some increase in available data in recent years.

Another technique of economic forecasting involves the use of econometric models.[2] These are mathematical and statistical descriptions of a business, economy, or industry. This approach is especially useful in long-range forecasts when many unknown factors are involved. By expressing these factors in equations, they can be easily varied and the effects of changes on the ultimate outcome can be assessed. In this way, many different assumptions can be used and the results examined; as a result, the most likely outcomes can be predicted as well as the range of possible outcomes.

Detailed descriptions of more sophisticated approaches, such as econometric models, are available in the literature as well as the information on more prosaic methods, including trend and regression analysis, time series analysis, extrapolation, and trend curves.[3]

Technological Forecasting

In recent years there has been a rapid increase in technological forecasting. New products and processes are becoming available to companies at ever-increasing rates. Despite this, relatively few companies are emphasizing regular and complete technological forecasts. The few that do place significant emphasis in this area generally come from high-technology enterprises. It has been said that the only certainty about the future is that it will be different from the present and the present will be different from the past. Determining just how much difference can be expected and defining ways in which these differences will be manifested are the primary objectives of the technological forecast.

Technological change is the result of economic, social, and political conditions. Consequently, the *first step* in making a technological forecast is to analyze whatever has already taken place for technology to arrive at this present state. The future can then be predicted based on where society seems to be headed. As an example based on social trends, let us consider the matter of air quality in our environment. This is an important subject today because of the emphasis on energy and especially on the utilization of coal in order to alleviate the need for oil importation.

A look at the past shows society's continuing interest in cleaning up the air. In 1941 in Pittsburgh, Pennsylvania, one could have a white sheet of paper covered with soot in a matter of five minutes, even if it lay on a desk. Since then, however, Pittsburgh has made tremendous strides in cleaning up its air through strict air quality standards placed on steel and other industries working in and around that city. This led to a complete change in the attitude of the people of the city and brought about the tearing down of old buildings and the construction of what is now known as the Golden Triangle. A review of technological change would indicate that significant strides were made in developing methods of handling the

[2]M. F. Elliott-Jones, *Economic Forecasts and Corporate Planning* (New York: The Conference Board, 1973).

[3]D. L. Hurwood, E. S. Grossman, and E. Bailey, *Sales Forecasting* (New York: The Conference Board, n.d.), Chap. 8.

effluent from the many industries at work in that city. The influence of social pressure and technology are both evident in these changes.

The *second step* in the technological forecast is to predict both the qualitative and quantitative changes in technology that are likely to take place in the future. Similarly, an understanding of social pressure is necessary. Taking the matter of air pollution again, the quantity of particulate matter and other contaminants such as sulphur discharged into the atmosphere by a particular industry in the year 2000 will be affected by several factors, including (1) changes in the amount of pollutants produced through the greater efforts to utilize coal, (2) changes in the number of plants that will have installed devices to control particulates and sulphur, and (3) the collection efficiency for particulate matter and the sulphur removal efficiency of the control systems in operation.

Very fundamental technical changes must be considered, such as the conversion of coal to liquids, removal of hydrogen sulphide during this process, and the transmission of these liquids to the point of consumption. Another change is coal gasification techniques with attendant sulphur and particulate removal, which may provide a clean fuel for boiler houses not requiring the large scrubbers which have been tested on the discharge gases from utility plants. Economic matters cannot be discounted in such analyses, since they will affect which processes may come into prominence and the attitude of society toward the utilization of these techniques, because the members of society will in the long run pay for the extra cost.

The *third step* is to reconcile the results of this second step with the requirements defined in the first step. In our example, society has continued to demand ever-higher standards in terms of gaseous effluent quality discharges as they affect air pollution. However, with the desire to use greater quantities of coal which contributes to both of these contaminants, a decision must be made on whether or not it is technically or economically feasible to maintain these standards. The government at the moment is talking about relaxing some of these emission standards in order to make it more feasible to utilize coal. The forecaster must then judge how society is likely to react during the two decades until the year 2000 in order to decide how technology will have to respond. He must bring together the economic, sociological, and technological matters in such a manner that they are compatible.

Of some value in these analyses is the matter of forecasting technological efficiency. If one were to plot efficiency of operation versus technological improvements over a period of years, he would obtain an S-shaped curve. This curve would indicate that efficiency improvements start out slowly, rise rapidly, and then taper off at the end of some period of time, say ten years. The tapering off is represented by how the efficiency improvement approaches 100% asymptotically (Figure 3–3).

The curve tells us basically that more efficient ways will be found for performing the same function. Take the case of the large scrubbers that have been placed on the discharge of flue gases from utility plants to remove sulphur-bearing compounds and particulate matter. The initial installations were not very efficient and probably were down being repaired more than in operation.

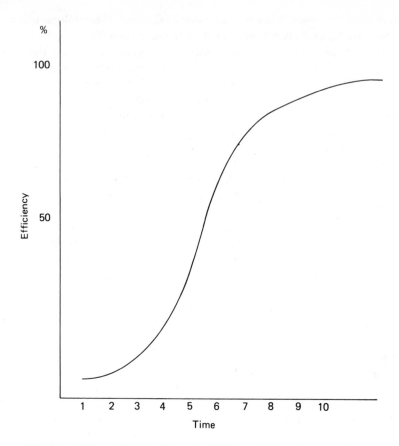

FIG. 3-3. S-Curve Showing Change in Efficiency of Operation Over Time

However, one can anticipate that if they follow the normal trends, their efficiency will perhaps increase to 80% or so at the end of a ten-year period. This is of value in technological forecasting in that if new techniques are being studied that are not particularly efficient, one can anticipate that modifications will arise which will increase the efficiency in the coming years. Sometimes major changes in equipment are necessary to bring about this efficiency. For example, originally dust collectors were mechanical in nature. Although some improvements were made to increase the efficiency of these mechanical systems, major changes did not occur until electrostatic precipitators came into being.

Several techniques have been used in technological forecasting. One commonly used method is known as the *Delphi technique.* It get its name from the oracle at Delphi in ancient Greece. The technique was developed in the early 1960s by Rand researchers N. C. Dolkey and O. Helmer. It is designed to improve the use of expert opinion through polling based on three conditions: anonymity, statistical display, and feedback of reasoning. In operation, a panel of experts are queried anonymously through formal questionnaires. The experts do not know who is on the panel and the replies are coded. Results come in to a central authority where they are averaged and the results fed back to the participants for comment, arguments, and counterarguments. The material is again returned to a central authority and the process is repeated through a number of iterations.[4]

Trend analysis is another method of predicting how future technology will be related to past performance. The trend involves modeling the past and then extending this into the future. There are various graphical and mathematical methods available for doing this. Among these are (1) linear regression, (2) nonlinear regression, (3) moving averages, and (4) exponential smoothing.

Another technique is sometimes known as *goal-oriented forecasting.*[5] This assumes that the technology will materialize to fulfill needs. It involves the discovery and evaluation of future needs and the evaluation of the likelihood that a specific area of technology will help to fill these needs. The analysis that results from this approach is interesting and useful to the planner, but it can really only be useful when it is combined with economic, social, and political trends.

Social and Political Forecasting

Today this does not receive much attention, but there are increasing signs that it will in the future. The signs include the problems of pollution, minorities, lawlessness, crime, and highway safety. It can be seen in the pressure to improve air pollution despite the fact that the country has done little to reduce its utilization of the automobile and to encourage mass transportation.

The entire matter of energy under discussion today is tied up in social attitudes. An appropriate solution will probably not be found until there is a suitable compromise between society's demands for energy and its willingness to accept changing standards of air and water pollution.

The political impact on corporations is immense and shall be discussed at greater length. From a forecasting standpoint, it is important to sense how Congress will pass new laws that will impact on the operation of the corporation.

The Overall Forecast

Forecasts in these three major areas will impact on engineer managers at various levels within the corporation. As more specific corporate operating plans are developed, these plans will center around the initial development of sales forecasts and their application in developing production requirements, inventory levels, and later the profit and loss of the corporation. They will impact on sales forecasts by indicating the environmental situation for business. They will impact on production plans through the availability and efficiency of technology. Engineer managers at all levels will be involved annually in providing some input to these forecasts or providing estimates of the impact of these forecasts on their particular spheres of operation.

[4]N. C. Dolkey and O. Helmer, "An Experimental Application of the Delphi Method to the Use of Experts," *Management Science,* (April 1963), 458–67.

[5]J. P. Martino, *Technological Forecasting for Decision Making* (New York: American Elsevier, 1972).

Impact of Government Regulations on Forecasting

No engineer manager of any organization can afford to disregard the activities and actions of the federal government in his forecasts. Today government has substantial influence on business through the corporate income tax structure, other taxing units, and its expenditures in the economy. Government expenditures account for approximately one-third of the nation's gross national product. These expenditures include a wide variety of goods and services which directly or indirectly affect every segment of society. With such activity, no level of government can be ignored in the short- or long-term planning of any business. In fact, the federal government is the most important single force influencing the firm's operations in the community, state, or country. For example, the federal government can and does assist certain areas of the economy through low-interest loans, tax incentives, and regulations. At the same time it penalizes other areas.

The government is a major consumer and employer. Furthermore, the federal government has the patronage of many powerful offices of the federal, state, and local governments. Thus, government can use various forms of pressure on groups throughout the business community to force them to conform to its policies.

If one doubts government influence, one must remember the argument that still continues in Britain: "to nationalize or not to nationalize." Today the United States government has all the control it "presently" needs through the process of collecting taxes, regulating certain organizations, and placing large contracts with industry.

The government collects revenues from society, and in turn it offers huge contracts to many large corporations. Their subcontractors reach into every segment of the economy. It is not just the profit from doing the work that attracts them; it is that the research and development pays off in commercial products for years to come. The federal government's contribution to research and development and supporting facilities has totaled almost $200 billion since World War II. Over one-half of this huge sum has been channeled through the Department of Defense. This contribution is just one way the government influences research and development; thus it has an extreme effect on the direction of the economy and society.

Today no business has complete freedom to decide its actions without considering the goals or objectives of society. Many engineer managers have learned that they must keep company policies and practices in tune with what the government and society believe "proper." To ignore society's goals leads to unfortunate results. Engineer managers must look to the future, to the changing economic and social climate, to the decline in demand for some products, to the rise of others, and to opportunities for new ones. But, they must also be alert to proposed legislation, must talk to senators and legislators, and read reports of cabinet members and other government officials.

During the course of the nation's history, society has continuously established laws and regulations to ensure that certain businesses do not harm society.

Among them are the Sherman Act, Clayton Act, Robinson–Patman Act, Federal Food, Drug, and Cosmetic Act, and acts establishing such regulatory commissions as the Securities and Exchange Commission, Federal Trade Commission, and the Federal Communications Commission.

The engineer manager must expect all levels of government to have an active role in all aspects of the business organization. Such government activity does not mean death to the free enterprise system. The engineer manager must be able, however, to anticipate with a fair degree of accuracy the effects that government activities will have on his particular organization in the future. To aid him, the government has steadily expanded and improved the data it makes available. This vast array of data can help him predict the economic environment in order to lay realistic future plans.

Over the decades society has charged the government with a number of responsibilities that were once assigned to individuals. Today society is interested in providing complete security for its members, e.g., increased welfare programs and housing. There are recent innovations in environmental and pollution control regulations to protect the health of its constituents. Large appropriations for defense-related items are now on the increase, and substantially affect the business environment. Changes in relationships with communist countries affect the export of both high technology and farm products. The engineer manager can ignore the influence of government only at his peril.

IMPLEMENTING PLANS

Having chosen the plan from among several alternatives, the engineer manager now comes to the important step of implementing the plan. In developing the plan it was necessary to set long-term goals or objectives for the organization. Manley Howe Jones has presented an interesting procedure that is useful in implementing plans and is known as the *means-end staircase*.[6] In operating his procedure, he sets three goals, which he terms *ultimate goals, intermediate goals,* and those *short-term goals* that rest on the steps of the staircase leading up to the intermediate goals.

He considered the first two goals as relatively permanent or far off. For example, the ultimate goal of a corporation might be to be recognized as the preeminent force in its particular industry. Intermediate goals are somewhat closer and somewhat more specific. For example, the intermediate goals of a company might be to develop new products or find a new group of customers which in turn would be a means for the company to use in order to establish its preeminence in its industry.

The third goal on the means-end staircase would represent that series of shorter-range goals that are necessary in order to develop a new product or secure the new group of customers.

[6]Manley H. Jones, *Executive Decision Making* (Homewood, Ill.: Richard D. Irwin, 1962), pp. 7–24.

Let us now turn our attention to the means-end staircase part of this concept. We will think of a means-end staircase as a set of steps on which rest statements of a man's or a company's lesser goals. Statements are arranged so that each serves as a means of attaining the goal resting on the step above it and as a goal when we are trying to decide on a means to place on the step just below it. We will visualize the topmost step as resting against one or more of the intermediate goals. The goal on this top step is a decision we have made on a means that we expect will be effective in achieving one or more of the intermediate goals we have chosen. When making each of these decisions we view it as a means, but thereafter we view it as a goal and in turn try to find a means of realizing it. Then, as soon as we can decide on this last-named means, we place it on the next step below that goal. This in turn becomes a goal, and so on down the staircase. Figure 3–4 will make this operation clear. Notice that the intermediate and ultimate goals are shown as part of the staircase.

Now let us look at how this might be utilized in the corporate example we mentioned. We have a corporation that wants to be the leader in its field and have at least one intermediate goal to help in achieving the establishment of a new product line. Now to achieve this new product line, we must have a means. Let us say that our means is to develop a new product that is needed within our industry. This, then, becoms a short-range goal. To achieve this goal, we need a means, and one of the means is to determine by a market survey what new product is needed within our industry. This market survey then becomes a goal, and the means to achieve this goal is to request the corporation's market research department to prepare such an analysis.

When the market research group develops such a product need, it can then become a goal for another set of staircases that proceed along the technical line. For example, the means for achieving this new product involves budgeting sufficient monies and allocating sufficient manpower to do research work to find a product satisfying these needs. This then becomes a goal, i.e., securing sufficient funds to support the research effort. This in turn requires a means of securing the funds. When this is achieved, then the goal for developing this product requires the necessary research personnel of proper quality to carry on the investigation. Providing appropriate personnel becomes a means of doing this, and in turn it becomes a goal, which is the hiring and allocation of sufficient personnel.

All of these means then become part and parcel of the implementation of plans. This form of thinking about plans is something we do not reduce to charts or to writing. It becomes a thought process that we go through to determine the various steps needed to achieve the goals or objectives that have been set. We likely will go through this thought process with lightning speed as we become more accustomed to its application.

It has certain other advantages in that it helps us find the steps that are missing in the total process and for identifying the subordinate staircases needed to achieve these short-range goals.

Success in achieving goals or objectives depends on finding the effective

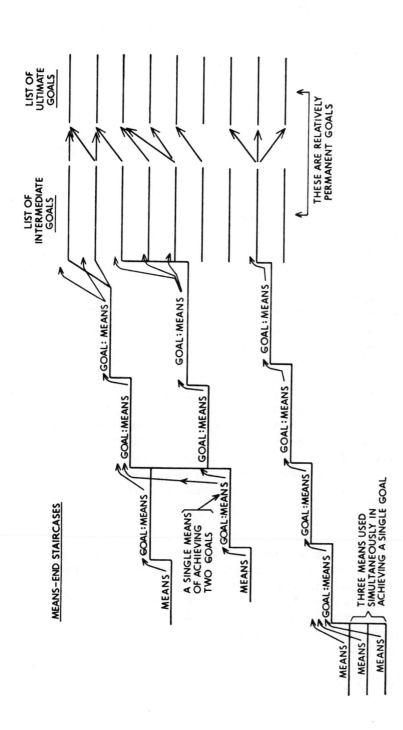

FIG. 3–4. Means-End Staircases and Permanent Goals [Source: Manley H. Jones, *Executive Decision Making*, rev. ed. (Homewood, Ill.: Richard D. Irwin, 1962) pp. 19 and 21. © 1962 by Richard D. Irwin, Inc.]

means which the engineer manager can employ. In a clearly framed means-end staircase, each step is an action that meets the following tests:

1. It is the means available for achieving the goal and the step above it.
2. It is a cause that is powerful enough to produce the goal on the step above it.
3. It serves as a goal for selecting an effective means to be placed on the step below it.
4. It is a decision that has not yet been carried out.

Several means-end staircases are customarily used simultaneously to attain the chief intermediate goals. These may be considered the subsystems shown in Figure 3–5. Certain of the steps contained in one staircase may serve as a means of furthering steps in other staircases. This was shown by the market research substaircase in our example.

The means-end staircase technique is a way of viewing decisions already made but not yet put into effect—in other words, plans that have not yet been implemented. By using this technique, the manager is able to reappraise earlier decisions and make sure that each decision is the most effective means at his disposal for achieving the goal just above it. Many times he may find that some of his decisions were weak or inadequate, or he may conclude that some of his decisions did not produce good results. He may conclude that a better alternative is now available, or he may find that an earlier decision has had some spurious effects. For example, he may have ordered a piece of equipment which in turn required hiring new people, who in turn were perhaps not adequately trained, and as a result, produced an inferior product. Perhaps he rushed onto the market a new product inadequately tested and a product liability suit resulted. Disciplined thinking through the goals and the means might have prevented these results.

PLANS IN OPERATION

The engineer manager, in carrying out plans, must set their courses so that they will be compatible with other activities within the entire organization. To effectively manage, he must make use of policies, procedures, and rules. *Policies* are the general statements or understandings that guide thinking and action in decision making. These may be verbal, written, or implied. Policies are used for consistency in various situations.

A policy is a broad, comprehensive statement. Therefore, it does not define exactly what to do; instead, it indicates the areas in which decisions are to be made (Figure 3–6). It helps to assure that decisions will be consistent with and

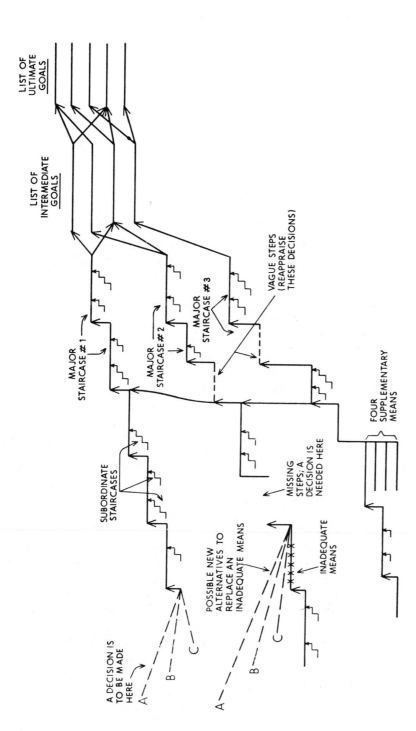

FIG. 3–5. A Fairly Complex Set of Means-End Staircases Showing the Points Where Decisions Will Have to be Made [Source: Manley H. Jones, *Executive Decision Making*, rev. ed. (Homewood, Ill.: Richard D. Irwin, 1962) pp. 19 and 21. © 1962 by Richard D. Irwin, Inc.]

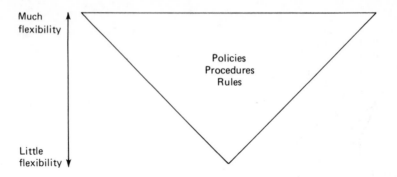

FIG. 3-6. The Comparison of Policies, Procedures, and Rules and Their
Degrees of Flexibility

contributive to objectives. A policy reveals the engineer manager's desires with respect to specific actions for future time periods, defines the general direction to be followed, and at the same time provides freedom to determine the appropriate action in keeping with the overall plans of the organization. These predetermined guidelines still provide freedom to determine the appropriate action in keeping with the overall plans of the organization of which the unit is a part. A policy does not make the decisions but delimits the area within which decisions are to be made. For example, departmental policies might be to hire only university-trained engineers, to have annual appraisals of performance, or to encourage employee suggestions, while company policies might include to promote from within, to provide two weeks' vacation for one year's service, or to seek business on a price basis.

Policies are a very important part of planning, for they provide the framework for the decisions that will be made to achieve the desired objectives. Policies tend to predicate issues, avoid repeated analysis, and give a unified structure to plans. This allows the manager to delegate authority while maintaining control. Consequently, good policies provide stability and confidence in the planning process by giving a steady course of action. In general, policies that are well thought out are infrequently changed.

Developing good policies cannot be overemphasized, for well-formulated policies aid the managerial process. Therefore, all aspects and consequences should be fully considered in the development of a policy. This will ensure flexibility under various situations. All policies should be reexamined and given special attention during the formation of a unit, department, or organized effort and during planning changes. In all cases, policies must be integrated so that uniform, orderly, and efficient execution of activities will complement one another. When this integration is lacking, misunderstanding and confusion are the result.

Good policies bring about desirable results for the engineer manager. It is important for him to recognize that the development and maintenance of policies require his attention. Therefore, periodic reviews of policies are necessary to determine if a given policy helps to get work accomplished. The engineer manager must answer the question, "Does this policy aid in improving the performance of work?" Because the organization is constantly changing, policies become obsolete and must be changed or discarded. Periodic reviews of policies provide an opportunity to determine if they are compatible, aid in achieving objectives, and integrated at each level.

A logical sequence in planning is to develop the objectives, the policies, and then the procedures. *Procedures* establish a standard method of handling future activities. They are guides to action, not to thinking, in that they detail the exact manner in which a certain activity is to be accomplished. They are specific in contrast to policies which define the general area in which the procedures will apply. Therefore, a policy requires interpretation by the engineer manager; a well-defined procedure requires no interpretation. A procedure should state the how, when, and who of an activity. Once the procedure is established, it can be used many times, thus relieving the engineer manager of determining a course of action each time work must be performed. In developing procedures, the engineer manager must use the same techniques followed in establishing policies and objectives. Examples of procedures would include handling of expense accounts, sick leave provisions, handling of orders, and purchasing of materials.

Procedures are frequently but not necessarily made up of rules. *Rules* are required actions which are chosen from among alternatives. They should not be confused with policies or procedures. A rule requires that a specific and definite action be taken or not taken with respect to a situation. Rules allow no descretion in their application. Examples of rules are "no smoking" in certain plant areas, wearing safety glasses in certain situations, or that all orders must be confirmed on the day of receipt.

To obtain satisfactory results from the planning process, the engineer manager should be aware of the following common pitfalls:

THE DO'S OF PLANNING

1. *Time commitment.* Planning requires a considerable amount of time; he cannot think of planning as a sideline.
2. *Timely planning.* Planning is never finished work, but plans must be completed before the activity starts. Even though the completed plans may be rough, they are preferable to no plans.
3. *Results expected.* Because planning takes time to implement, he cannot expect immediate results.
4. *Considered consequences.* Increasing sales is always a goal, but he must consider the effects on production, marketing, and profitability.
5. *Obtained information.* It is impossible to obtain all of the information desired; therefore, he must at times make assumptions.

6. *Participation in planning.* The manager must involve others in planning, but he must provide the leadership. Outsiders can provide the steps and offer suggestions, but they cannot formulate the plans.

7. *Comprehensive planning.* Planning is a comprehensive process, and separate or independent plans cannot be tolerated.

8. *Communication and understanding.* Plans must be communicated, not kept in a desk drawer.

SUMMARY   The purpose of every plan is to facilitate the accomplishment of objectives. It is the function of all managers, although its character and breadth will vary with their authority and the nature of the policy and plans outlined by their superiors.

The planning process is made up of a number of steps, including recognizing opportunity, setting objectives, developing information, developing alternative plans, and selecting the plan. Each of these must be done carefully and in an orderly manner using the best available information, including forecasts. These forecasts will include economic, technological, social, and political forecasts.

Economic forecasts will utilize certain information published at the national and regional level dealing with the gross national product, demographic changes, and employment. Technological forecasts are made based on analyses of past technological trends and the likelihood of future technological improvements brought about by social pressures and research and development. Various analytical methods, including the Delphi method, are used in forecasting these technological trends. Social and political forecasts are important because of the impact they have on economic and technological matters.

The importance of the government cannot be discounted in forecasting as well as in operation. Government permeates every segment of society. Because of its manpower requirements and its massive purchasing ability and its regulatory agencies, the future of industry is dominated by its influence.

Plans must be implemented effectively. The means-end staircase system provides an interesting way to review decisions that are to be put into effect and to provide proper implementation.

When plans are once in operation, policies, procedures, and rules become important matters. They are important in order to maintain uniformity of operations as well as to allow the assignment of responsibility and delegation of authority in the decision-making process.

Planning is a time-consuming process but an invaluable one. Unless one sets up an orderly procedure to achieve objectives, he cannot operate his department or corporation effectively; in the difficult competitive arena of today's society he will be a failure.

IMPORTANT TERMS

Goals: The operation toward which an effort is directed.

Policies: General statements or understandings which guide thinking and actions in decision making.

Planning: The setting out in an orderly fashion those steps necessary to achieve a goal.

Procedures: Guides to action which provide a standard method of handling future activities.

FOR DISCUSSION

1. Explain the difference between policies and procedures.
2. What are some of the common pitfalls in planning?
3. The planning process requires the acquisition of relevant information. What are sources of information from both within and without the company to aid this process?
4. What is the impact of new technology in the planning process?
5. Discuss the steps of the planning process.
6. Why are alternative plans important?
7. What is the means-end staircase?
8. Why is the means-end staircase a useful device for the implementation of plans?
9. Describe your ultimate personal goal, your intermediate goal, and several of the means-end staircases that you will utilize in reaching your intermediate and ultimate goals.
10. Describe the ways in which the government influences business operations today.
11. What forecasts are of importance in developing an overall corporate forecast?
12. What techniques are useful in the development of forecasts?
13. Describe the Delphi technique.
14. Why are "policies the framework for decisions"?

CASE 3–1

Planning the Overtime

Tom Johnson, Vice President of the Metal Products Company, called Rex Smith, Chief Engineer, to his office. "Rex," he said, "this excessive overtime has got to stop. You have already used all your budgeted overtime and the month is not even half over. What is the trouble in your department?"

"Tom, you will recall when pressure was on to reduce costs, I transferred two engineers over to the Belleville plant. According to the sales forecast, we would have been able to handle all project designs, specifications, and bids required without working overtime. Now the sales department is moving the delivery date on orders and increasing the number. Right now I don't have as much help as we need to prepare the work for manufacturing as fast as sales is calling for it." Also, two of our best engineers have been sick and another one has been on vacation.

"Rex, I've told you time and again not to accept schedule changes from sales without first making sure you can handle them. It is easy for sales to make promises, but the way you're running up overtime, we are the ones who will pay for those promises."

"Yes, Tom, I understand, but I really believe we will be able to handle these changed schedules with very little more overtime and be in good shape by the end of the month. Besides, one of the major goals of the company is to make deliveries on time and serve the customer."

"Rex, I still don't think you are handling this problem very well. When costs go up and profits go down, sales will blame us!"

A week later Rex's department is still behind and the sales manager calls. "Rex, my key salesman tells me you are holding up the North American order. They need the drawings and bids by tomorrow. Can I tell my people you will have them ready by noon tomorrow?"

1. What should Rex tell the sales manager?
2. What do you think of Rex's instructions from his boss?
3. If you were Rex, how would you handle the problem?
4. How can Rex avoid future problems of this type?

CASE 3-2

Proper Communication

For a number of years the manufacturing department of the Metal Products Company had operated with a high level of productivity, enthusiasm, and morale. People seemed to work fast and with a minimum of lost time.

However, during the current year problems seemed to be developing which were causing errors that required work to be discarded or redone. Tom Johnson felt that this was probably due to a breakdown in coordination or perhaps some personal conflicts. He had some analyses made on the situation and reached several conclusions. First, he felt that the corporate level managers were not communicating their goals and policies clearly to the operating level managers. In turn these managers felt the top management was planning and communicating in a vacuum. They were making plans without relevant and realistic information from the operating levels. Lower-level managers in manufacturing and marketing frequently found themselves working at cross purposes and with little coordination from above.

Second, frequently key people at the operating levels failed to "get the word" at all, or they got it indirectly through informal channels, perhaps too late.

Third, there were many cases of other functional departments at the operating level misunderstanding one another. At one time the personnel department had performed a study for the sales department. However, when the sales department reviewed the study, it found that it was not really what it wanted and promptly filed the study away. Then one of the manufacturing sections felt that it needed some computer service, but what it got really did not do the job. There seemed to be a lack of a common level of communication.

1. What is Tom's problem?
2. How is the planning process breaking down?
3. Is there a communication failure?

FOR FURTHER READING

ARGENTI, J., *Systematic Corporate Planning*. New York: John Wiley & Sons, 1974.

AYRES, R. U., *Technological Forecasting and Long-Range Planning*. New York: McGraw-Hill Book Company, 1969.

BOX, G. E. P., and G. H. JENKINS, *Time Series Analysis Forecasting and Control*. San Francisco: Holden-Day, 1970.

DRUCKER, PETER, *Management*. New York: Harper & Row Publishers, Inc., 1974.

EILON, S., "Goals and Constraints," *Journal of Management Studies* (October 3, 1971), 292–303.

EWING, D. W., *Long-Range Planning for Management,* 3rd ed. New York: Harper & Row Publishers, Inc., 1972.

FAYOL, HENRI, *General and Industrial Management*. New York: Pitman Publishing Corp., 1949.

FULMER, R. M., and L. W. RUE, *The Practice and Profitability of Long-Range Planning*. Oxford, Ohio: The Planning Executives Institute, 1973.

HERTZ, D. B., "Risk Analysis in Capital Investment," *Harvard Business Review* (January–February 1964), 94–106.

KAHN, H., and B. BRUCE-BRIGGS, *Things to Come*. New York: Macmillan, Inc., 1972.

KOONTZ, HAROLD, "Making Strategies Planning Work," *Business Horizons* (April 1976), 37–47.

ROSS, J. E., and M. J. KAMI, *Corporate Management in Crisis: Why the Mighty Fall.* Englewood Cliffs, N.J.: Prentice-Hall, Inc., 1973.

SCHAEFFLER, SIDNEY, et al., "Impact of Strategic Planning on Profit Performance," *Harvard Business Review* (March–April 1974), 137–145.

SCHAFFER, ROBERT, "Putting Action into Planning," *Harvard Business Review* (November–December 1967), 158–167.

SULLIVAN, W. G., and W. W. CLAYCOMBE, *Fundamentals of Forecasting.* Reston, Va.: Reston Publishing Company, Inc., 1977.

WALL, JERRY, "What the Competition Is Doing: Your Need to Know," *Harvard Business Review* (November–December 1974), 22–41.

WARREN, E. K., *Long-Range Planning: The Executive Viewpoint.* Englewood Cliffs, N.J.: Prentice-Hall, Inc., 1966.

WHEELWRIGHT, STEVEN, "Strategic Planning in the Small Business," *Business Horizons* (August 1971), 51–58.

WORTMAN, M. S., and F. LUTHANS, *Emerging Concepts in Management,* 2nd ed. New York: Macmillan, Inc., 1975.

THE TRANSITION TO ORGANIZING TECHNICAL ACTIVITIES

The engineer manager must establish an organization to achieve the objectives under his responsibility as they relate to the corporate objectives. He must learn to work through both formal and informal relationships within the organizational structure.

LEARNING OBJECTIVES

- ☐ Gain appreciation of the concept of organizing.
- ☐ Learn how to develop organizational charts.
- ☐ Understand the operation of line and staff organizations.
- ☐ Appreciate the need to use committees effectively.
- ☐ Learn to cope with organizational changes.

EXECUTIVE COMMENT

JAMES F. BOWEN
General Manager, Switched Services
Southwestern Bell Telephone Company

James F. Bowen joined the Bell System in 1954. Since then he has held numerous positions in AT&T, Long Lines, and in the Plant, Engineering, and Marketing Departments until his promotion to his current position of general manager—Switched Services.

Organizing Technical Activities

One of the ongoing requirements of an engineer manager is to be able to organize the technical activities for which he is responsible. The driving force behind this requirement is the manager's personal commitment to the goals and objectives of his part of the action plus his contribution to the overall goals and objectives of the entity, corporate or otherwise, with whom he has chosen a career path.

Perhaps for the first time comes the realization that the managing of those technical activities requires skills that are not covered by the many laws of engineering science. Also, for the first time, some undefined elements of risk appear which must be woven into the organizational model. The human factors present in all organizations assure this.

In my organization I like to think of this as a "sense of proprietorship," that is, each manager organizes his part as if he were the owner. To do this well, each must have a knowledge of various forms of organizational structure, strong and weak points, and situational effectiveness. Perhaps the greatest single challenge for the manager is to be able to change the organization when necessary and to effectively cope with the changes of others with whom he interfaces.

Many managers, recognizing the unique background and educational training of their technical staffs, adopt a *participative management* style. If this is done well, the organization as a whole is contributing to the organizing, managing, and doing of the effort necessary to reach their goals and objectives. In any organized system this management style appears to mold the best of the formal and informal organizations into a permanent blend that, coupled with a cultivated "sense of proprietorship" on the part of all managers, produces a very strong commitment to the success of the group as a whole. I have seen this work well in my organization and recommend that it be given serious consideration.

The engineer manager who effectively organizes his subordinates and other resources increases the likelihood that he will achieve the goals and objectives within his area of responsibility. Such organizing requires him to work through formal and informal relationships both among members of his group and with other groups.

Organizing involves defining activities that are necessary to achieve planned goals or objectives, obtaining resources, establishing informal and reporting relationships among people, and allocating planned subgoals to those people. These activities bear some similarity to those which the engineer manager pursued as an engineer. Then, however, he was concerned primarily with physical things. His concern was to define relationships between parts of a physical or chemical system such that those parts would interact in a certain way. As part of the design, he made sure that a system would not only operate properly but that it could also be adjusted if necessary. Such design work is so much a part of the routine for the engineer that he can design a system without realizing the similarity to the managerial task of organizing.

Organizing for the engineer manager means defining a properly operating and adjustable system of relationships among subordinates and the resources available to them. Like an engineering design of a system, an effective organization is one that contributes to the achievement of objectives. Unlike an engineering design, the organizing of the activities of subordinates requires giving them a degree of independence. That latitude is often complicated by the unpredictability of subordinates' behavior. Such complications can loom so large in the mind of the engineer manager that he either overlooks the need to organize resources or concentrates on the resources at the expense of people organization.

An effective organizational structure will provide a means for people and resources to function as a system. Moreover, the structure will complement the directives of the engineer manager under which each subordinate is held responsible for certain specific results. The organizational structure removes a large amount of uncertainty and confusion that would otherwise exist within the group.

REASONS FOR ORGANIZING

Organizing is the first step toward making plans effective. It is also the guide to staffing. Organizing requires the engineer manager to divide work into meaningful and manageable "units" and assign each unit to specific groups or individuals.

Such assignments achieve both the advantages and the disadvantages of specialization. For example, there is always the question of which subordinate gets which units. The more units a subordinate can effectively handle, the better, and there are no predetermined maxima or minima. In practice, however, the more work units a subordinate handles, the more contacts he will need with the manager. Nevertheless, from the viewpoint of time economy, the manager needs to control the frequency and duration of direct contacts with subordinates. Some of the factors that affect the time spent in personal contacts and over which the engineer manager has some control are as follows:

1. *Planning.* The more carefully the engineer manager defines plans, schedules, and objectives and relates each to his subordinates, the more likely they are to know what he expects of them and the longer they can operate without personal contacts, particularly those of an emergency nature. As explained in Chapter 3, planning involves developing policies, procedures, and rules, as well as following through by communicating the plans to subordinates.

2. *Training.* Better trained subordinates require less time to perform assigned tasks and fewer contacts with their manager. Effective training substitutes a large initial investment of time for the even larger amount of time required by a series of personal contacts to correct errors brought on by deficiency in training.

3. *Communication.* The demands of thorough communication create a dilemma for the engineer manager. On one hand, subordinates must receive any and all information that affects their function in the organization. On the other hand, such transmissions require personal contact by the manager and thereby limit his span of control. He can enlarge that span by requiring more written communication on his part and on that of his subordinates. Span of control can also be larger if both manager and subordinates more carefully plan and schedule their oral communications. (For a more thorough treatment of communication, refer to Chapter 11.)

4. *Delegation.* When the engineer manager clearly delegates well-defined tasks, the well-trained subordinate is able to accomplish these tasks with few additional instructions and attention. (The techniques of effective delegation are treated in Chapter 6.)

5. *Evaluation.* Personal contacts intended primarily as follow-up will show how closely subordinates are following plans and if they need help. Such contacts will be more effective if they are included in the original assignment, if they deal primarily in exceptions, and if they cause a minimum amount of interference with the subordinate's work schedule.

By implementing these techniques, the engineer manager can realize the cost savings and organizational simplicity of a larger span of control, and his subordinates can become more productive. Although there is a limit to the number of persons a given manager can effectively control, the exact number varies less with inherent personal factors and more with the manager's ability to use these techniques.

DEVELOPING THE FORMAL ORGANIZATION

In organizing, the engineer manager is concerned with balancing such factors as distribution of work to individuals, physical resources required, relationship of work units, and who is actually involved in the performance of the work. Forming an organizational structure to accomplish the work to be done requires that the engineer manager first break the work down into units to allow specialization and to permit participation by several subordinates. Second, the manager, having considered the circumstances of the firm, the abilities of his people, the workplaces, and the resources available, assigns the work units.

Often, decisions about activities depend on the available people and their particular interests and capacities. As a result, the engineer manager finds that assignments frequently overlap; more than one person is responsible for a given task, or some tasks are left undone. Symptoms of an ineffective organization include indecision, buck-passing, and a constant state of crisis.

Assignment of Personnel

The assignment of personnel to units within a logical and well-defined organizational structure is invaluable in assisting the engineer manager to achieve the goals of his unit. There are a number of basic techniques of grouping activities available to the engineer manager. In practice, combinations and modifications of the following techniques will fit particular situations:

1. *Grouping by numbers* is ideal when the amount of work done depends on manpower and when a given number of persons are managed by one individual. This method of grouping is used in such situations as common labor crews.

2. *Grouping by functions,* such as production, research, development, sales, or finance, often is a result of the existence of those functions from the early days of the company.

3. *Grouping by geographical areas* means that all activities in a given area are the responsibility of one individual.

4. *Grouping by product* causes supporting activities to be assigned to one individual. This method is used when rather unrelated product lines exist within a single firm. Each product group or department has subgroups for production, research and development, marketing, and engineering. Such organizational structures allow individual attention to related product lines, thus determining more accurately the profitability of the line.

5. *Grouping by customers* allows the firm to provide specialized services that can be specifically identified. This type of grouping is used by companies that sell to several industries and government agencies.

6. *Grouping by process* allows one individual to be responsible for a particular manufacturing operation, e.g., stamping, assembly, or painting.

An effective decision about which formal grouping to use involves consideration of the strengths and weaknesses of each of the six alternatives. No matter which basis the manager chooses, he cannot expect well-conceived, rational relationships to evolve automatically; they need his personal attention. After making these decisions, the engineer manager is now prepared to develop an organization chart.

The Organization Chart

The organization chart shows reporting relationships. Its first draft will often show inconsistencies and complexities that can be corrected. An organization chart also defines the major activities and their respective relationships. Pertinent supplementary information in connection with the chart may be included in an organization manual.

An organization chart is customarily arranged so that major authority occurs at the top, with successive lower levels having less authority (Figure 4–1). Lines connecting the various blocks which represent activities show the channels of formal authority. Generally, the engineer manager will develop several organization charts. One will show the overall organizational structure and the relationship to other departments or units. Another will show the details of major components of the department or unit. Positions shown on the same horizontal level usually are considered to have the same relative importance, regardless of the level to which they report.

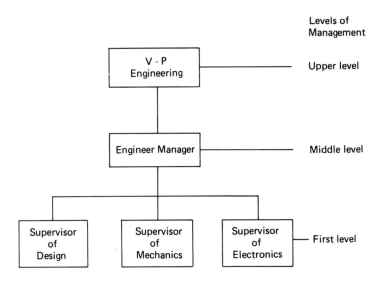

FIG. 4–1. Typical Organization Chart with Designated Levels

Most intergroups and interpersonal relationships in a firm do not appear on the formal organization chart. Their invisibility, however, requires an even more thorough understanding for effective managing.

INFORMAL GROUP RELATIONSHIPS

   Informal group relationships exist wherever there are people. Such relationships increase in importance when groups must cooperate in the attainment of planned goals and objectives. Therefore, it is essential that the engineer manager make certain that the members of those groups are familiar with planned goals or objectives. Such communications will be more effective if they are given on a need-to-know basis without regard to apparent levels of importance that may be inferred from the geometry of the organization chart.

   To be effective, the engineer manager must know the dynamics and leadership of the group and the identities and motivations of the people involved. Such knowledge will help him determine the group's influence and how to utilize the group effectively rather than resist it. He should determine why the relationship developed and what maintains it (mutual interests, physical proximity, or unique working conditions). The existence of strong informal groups may mean that the formal organization is obsolete. On the other hand, no amount of changes in the formal structure will eliminate informal group relationships.

   Should the engineer manager identify adverse consequences of the activities of an informal group, he will usually be better served by reducing the causes of those activities rather than by trying to destroy the group. Alternatively, he can seek constructive channels for the activities of the group.

Organizations are classified according to the nature of the reporting of relationships which can be identified as line, functional, or line and staff. The type of relationship is determined by the functions of groups and the assignments of individuals. Generally, no one type is used to the exclusion of all others. On the contrary, the realities of practice require that some combination of all three or modifications of them be used to enable the engineer to manage human and physical resources in a particular situation.

Line Organization

A line organization is the simplest form of organization. There is a direct superior–subordinate relationship (Figure 4–2) in which direction flows downward and accountability flows upward. The smaller the firm, the more likely it is to have a line organization. Service-type operations also usually have a line organization.

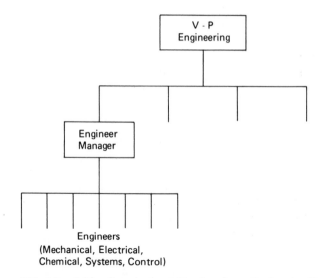

FIG. 4-2. A Line Organizational Structure for an Engineering Function

Functional Organization

The functional organization was developed by Frederick W. Taylor.[1] He could not obtain employees qualified to perform the many special technical managerial tasks required of his supervisors in the traditional line arrangements. To overcome this limitation, he divided the various technical tasks among his subordinates so that each would become better qualified and more of a specialist in his area. This division of expertise required workers to take directions affecting a given technical discipline from the individual responsible for that discipline in the

[1]F. W. Taylor, *Scientific Management* (New York: Harper & Row Publishers, Inc., 1947).

organization. A worker under this organizational arrangement has more than one superior.

Figure 4–3 shows such a functional organization. The specialists report to the engineer manager. An engineer to whom the engineer manager has delegated tasks takes direction from the specialists whenever the engineer is involved in their areas of expertise. The functional organization structure has seen only limited use in industrial organizations. It is used in small manufacturing firms and in new departments just starting operations, which then later change to a more appropriate organizational structure.

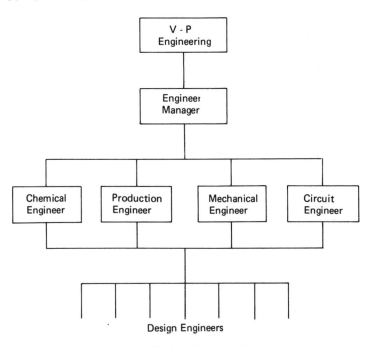

FIG. 4–3.   A Functional Organizational Structure for an Engineering Function as Suggested by F. W. Taylor

Line and Staff Organization

Line and staff organizations take advantage of Taylor's functional features and still have most of the advantages of line structures. These hybrids are the most common type today (Figure 4–4). Essentially, line supervisors are accountable for the operations and the specialists (staff) are accountable for providing sound advice and assistance when requested by the line supervisors. Line people have overall responsibility for achieving the planned goals and objectives. This organizational structure provides and uses people with specialized knowledge, thereby relieving the line people of the requirement of knowing everything. Generally, it is well to start with a line structure and add staff specialists as the organization expands.

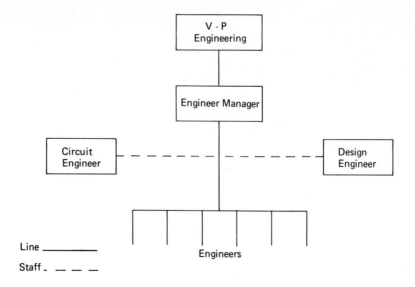

FIG. 4-4. A Line and Staff Organizational Structure for an Engineering Function

The advantages and disadvantages of the three types of organization can be summarized as follows:

1. *Line*
  A. Advantages
  (1) Definite supervisor–subordinate relationship throughout organizational levels
  (2) Easy to develop and understand
  (3) Subordinates have only one superior
  B. Disadvantages
  (1) Difficult to justify hiring specialists for technical expertise
  (2) Supervisor must acquire broad range of skills
  (3) Supervisor must be concerned with day-to-day activities

2. *Functional*
  A. Advantages
  (1) Allows for the development of specialists in the organization
  (2) Employees can obtain counsel from specialists
  B. Disadvantages
  (1) Subordinates have more than one supervisor
  (2) Lack of overall leadership; no one supervisor has overall accountability
  (3) Allows for conflicts to develop between specialists and line supervisors and among specialists

3. *Line and Staff*
   A. Advantages
      (1) Allows for specialists in the organization
      (2) Line supervisors responsible for major functions
      (3) Subordinates know their supervisor but still have staff specialists available for assistance
   B. Disadvantages
      (1) Decision making delayed because line must obtain advice from staff
      (2) Since staff not aware of line problems, their advice may be weak and inapplicable

Effective performance by both line and staff managers is essential to achieving the maximum capabilities of an organization. In the experience of most engineer managers, however, there is no area that causes more difficulties, problems, and friction than line and staff relationships.

LINE AND STAFF RELATIONSHIPS

The first step in reducing the friction and having an effective line and staff relationship is for the engineer manager to understand line and staff roles. The second step is to communicate that understanding to everyone in the organizational unit.

The question that immediately arises is what is line and what is staff. An appropriate definition is that the line position has direct supervision over the accomplishment of work to achieve the planned objectives and staff positions provide assistance and services to line positions in achieving these objectives. Ideally, staff services and assistance are rendered only when requested by line people. The word is "requested," not "forced." Although line-type superior-subordinate relationships exist within staff departments, the primary activities of the staff are advisory in nature. Therefore, the most important quality of staff employees is their ability to relate to line people. This distinction can imply that line managers are expected primarily "to act" and staff people are "to think and advise." The engineer manager's subordinates should know if they are being held accountable for acting in a staff or line capacity.

Effective Line and Staff Relationships

One way to ensure effective line and staff relationships is for the engineer manager to develop his staff through training programs to provide information and services that fulfill the needs of line operations. Because staff units must co-operate and work with others, the people assigned to them should be cooperative and helpful, either by nature or by training. Another way for the engineer manager to obtain effective relationships is to avoid duplicating expertise in a staff unit that already exists in the line units. He needs to foster an atmosphere in which line and staff are combining their talents in working toward common objectives.

Line units can benefit by examples of how to take advantage of advice from staff people. Such examples can come from the engineer manager who demonstrates in his own work how to utilize staff units effectively. As a rule, his line people get deeply involved in day-to-day operations and must be shown how and encouraged to utilize staff resources effectively. They should not just be told simply that staff people are available and "if you need them you may call them." Under this type of direction, staff resources probably will never be used by the line.

Specialized Authority

Often the engineer manager may find that forces within the firm make it necessary and advantageous to give staff units in the organization specialized authority over specified operations, practices, and activities relating to work performed in a line department (Figure 4–5). Some of the reasons that the granting of such authority is beneficial are as follows:

1. It ensures that all groups follow a uniform policy. For example, the accounting department has specialized authority to demand that the same depreciation schedules and standards of accuracy are used throughout the firm and the quality control unit has specialized authority to enforce the policy of not shipping products that do not meet given specifications by rejecting shipments.

2. Many times line managers do not have the time, ability, or interest to supervise a special program. Therefore, staff people carry out the program so that it receives the proper attention and direction.

There are both outside and inside forces that require specialized authority. The only sound basis for delegating specialized authority to a staff person within an organized unit is the expertise or knowledge of the individual. Specialized authority may get out of hand unless it is carefully restricted to the areas of staff expertise. The main outside forces that cause the establishment of such specialized authority are government agencies and labor unions. Relations with both require expertise in a given area which is frequently more effective when managed uniformly throughout the organization. The engineer manager relies on specialized authority sparingly and only when a real need exists.

USING COMMITTEES EFFECTIVELY   Committees and meetings are inevitable. It is not inevitable, however, that a committee be a source of frustration or a means of wasting time. It only seems that way. A thorough understanding of how committees function and the use of techniques for effective committee operations can avoid most of the problems. In fact, the committee can become an effective means of accomplishing organizational goals or objectives.

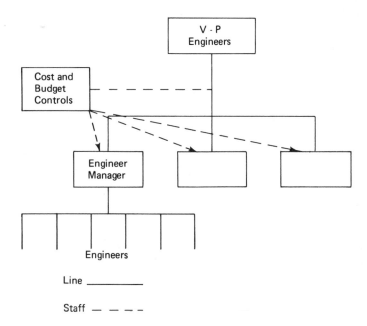

FIG. 4-5.   Line and Staff Organizational Structure Showing Functional or Specialized Authority of Staff

A committee consists of individuals who are appointed or elected to it. A committee may be chartered to deal with one subject, a few subjects, or any subject that is brought before it. The reasons that committees have such wide latitude and are used so often include the following:

1. *The collective knowledge or expertise of a group* can focus on specific problems. The committee acts as a forum to interchange and examine a wide range of ideas that would otherwise go unheard. New ideas can come to light and develop a measure of maturity. Clarification of problems can suggest novel solutions.

There are times, however, when ideas, judgments, and solutions can be obtained more effectively through individual conferences than through a committee meeting. The benefits of a committee can be obtained by having someone contact persons individually to obtain ideas and recommendations without actually forming a committee. Such contacts avoid the long deliberations of committees and are especially feasible in one-time situations.

2. A meeting can be a *means of disseminating information simultaneously* to all affected individuals. Instructions, decisions, and an opportunity to ask questions can be given to everyone at once. The attitudes and receptiveness of the participants are sometimes obvious. Emphasis of certain points can be made by gestures and other nonverbal means.

3. *Sharing of accountability and combining of delegated authority* among individuals in the organization are both possible through a committee. Sharing may occur when one wants to take full accountability for making a decision. Combining authorities to make a decision commits several persons to the decision. Many times a committee is used to choose bonus programs or to authorize major capital investments because of the unwillingness of a chief executive to give the authority to make such decisions to any one subordinate. There are times when several groups of people need to agree to make policy changes that affect all groups but no single group has the desire to make a decision that affects the other groups. A committee consisting of representatives of all groups may be able to break the impasse.

4. *Participation in the decision-making process* by members of a committee makes it more likely that the members will carry out a program. In its role as a forum, a committee lets interested groups be represented in the decision-making process. By this method, integrated planning can be achieved. The individuals concerned can be involved in developing overall plans and can know what each unit in the organization will be expected to do. A committee is especially useful when employees or groups complain that they feel outside the decision-making process. By being members or represented on the committee, they are more likely to accept the results.

5. Committees can be *used to avoid making decisions.* Many times committees are appointed to ensure either that no decision will be made or that the decision will be delayed. A symptom of such a mentality is the appointment of a committee with the sole charge to study the problem. Study committees are an inefficient use of resources. Once the members realize what has happened, they will be understandably uncooperative in future committee assignments.

Evaluation of a Committee's Performance

Because of the above reasons, direct or indirect involvement with committees is an inescapable aspect of managing. If for no other reason than to save managerial time, every effort should be made to ensure the effective use of committees. Such an effort requires evaluation of the following factors on a regular basis:

1. *Role* of committees. A committee cannot do everything. Every problem cannot and should not involve a committee. On the other hand, committees can develop major goals, objectives, and policies.

2. *Responsibility* of committee members. Each member of the committee should know his responsibility. Some of that knowledge will come from defining the scope of subjects the committee is expected to consider. Committee members can waste time while differing on what each of them is to do.

3. *Membership* of committees. Differences in ability, capacity, and communication skills mean varying levels of effectiveness among committee members. Such variations affect how well members represent their own group and the quality of their contributions to decisions that are in the interest of the organization. If the members are on the same level in the organization, constructive disagreement among members will be more effective. Only persons who will take the time to attend meetings should be asked to serve.

4. *Size* of committees. Make a committee too large and it may never reach a consensus on anything. Furthermore, the greater time required to allow all members to participate in the meetings will mean longer meetings and fewer accomplishments. If many interests must be represented, the selection of members becomes even more important. The need for representation may be overemphasized in many situations because the function of a committee is to evaluate the various aspects of the situation as a whole, rather than to protect specific interests. Committees may range in size from 6 to 12, with a good rule being "the smaller the number the better."

5. *Chairperson.* Much of what a committee accomplishes depends on the chair. Unfortunately, the chair of an important committee will often be given to someone more in recognition of his rank or his interest in obtaining power than his ability to get a diverse membership to cooperate. An effective chairperson will:

A. Plan the meeting and prepare for questions that may be raised.
B. Familiarize himself with the techniques to be used in conducting a meeting.
C. A few days in advance of the meeting send out notices giving the time and place of the meeting, the items to be covered, appropriate material or research data, and the approximate duration of the meeting. He may ask each member to call him to verify receipt of the notice.
D. State on the notice his expectations of each member to, for example, supply advice, arrive at a decision or a solution to some problem, or vote on a certain proposal to be presented.
E. During the meeting keep the discussion on the subject. Get all members involved in the discussion and in making preparation for future meetings.
F. Not permit telephone calls to be transferred to the meeting room while the committee is in session.
G. Ensure that minutes of the meeting are distributed within one working day after the meeting.

When the chairperson performs the above, the committee's efforts will probably be productive and worthwhile. The effects of the chairperson's performance are, however, even broader. The attitude of the chair affects the attitude of the members. If the chair is enthusiastic about what the committee is doing, then the members will want to be involved and active. If the chair is hesitant or unfamiliar with the problems, then one of the members may dominate the discussion. The other members will either rebel or mentally withdraw from the meeting. In either case, the meeting can be considered out of control and a time waster for other members.

When a committee fails to meet its goals, the failure is usually due to an incompetent chairperson.

6. *Evaluation* of a committee's work. Evaluation of the performance of a committee can determine if the committee is worth its cost. Costs are primarily in man–hours and are often much larger than management would have predicted. Time costs are easy to measure, but the benefits of a committee are generally more difficult to determine. This difficulty, however, justifies the effort to put an economic value on their efforts. The economic value of their objectives is most fruitfully estimated before the committee is constituted.

Ideally, the charge to a committee will include the expectation that its work must be completed by a given date. Omission of the time factor can mean that committees become self-perpetuating. Committees that complete their work on time deserve to be informed that their recommendations were used, or if not, why not. Any constructive action by the committee merits an expression of managerial appreciation to the members.

CHANGING ORGANIZATIONS

An organizational structure that may have taken months to initiate and develop will not last forever. It is understandable that the considerable amount of work of dividing activities into logical units, assigning people, acquiring necessary resources, and perfecting an organizational structure would justify the feeling that the organizing phase was complete. The organization, however, is constantly changing. Internally, people are acquiring new skills and knowledge; new people are being added to the organization; new equipment, products, and techniques are being added to the operation. New laws and regulations, competition, and technology are appearing. These factors require changes in the structure of the organization to provide new life and growth (Figure 4–6). Unfortunately, such changes suggest that previous changes that seemed very logical and correct when made may have been mistakes.

A complete annual reevaluation of an organization will identify areas in which changes should be made. Needs for change can be detected by answering the questions in Table 4–1.

Coping with Organizational Changes

The more necessary a change, the more difficult is its undertaking. Therefore, the changes needed most are postponed until they cause major problems. Changes that involve physical resources are relatively easy to make and generally are taken care of easily, whereas changes that involve people cause infinite problems. A job for which an individual was very well qualified and for which he received satisfactory performance ratings five years ago may be beyond him today. Further, the individual may resist even the suggestion that his assignment should be changed.

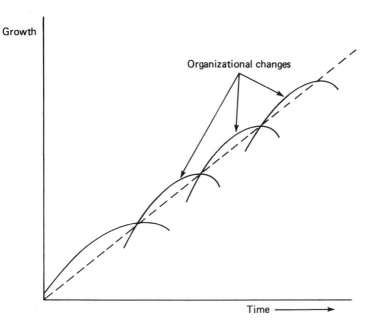

FIG. 4-6. An Organizational Life Cycle Is Characterized by a Series of Organizational Changes Which Propel the Organization to New Life and Growth

---

TABLE 4-1

Is Your Department in Organizational Trouble?

If your answer is "yes" to four or more of the following questions, the organization can probably be improved.

1. When a position is vacated, does there seem to be an excessive amount of time and money spent in finding a replacement?
2. Is more than 10% of an engineer manager's time spent in committee meetings?
3. Does as much as 1% of the total company salary expense go to "coordinators" and "assistants to"?
4. Are excessive organizational levels involved?
5. Do more than ten people report directly to the engineer manager?
6. Has the idea that the organization can be improved been rejected?
7. Does there seem to be a tendency for the engineer managers to do their important tasks themselves so that they will be done correctly?

Any decision resulting in organizational change must consider the effect such action will have on the people involved, their future performance, and their career with the company. Otherwise, the change will generate more problems than it will solve.

Some techniques that may correct people-related organizational problems are as follows:

1. *Training and counseling* to improve performance are especially effective in assisting individuals to assume additional responsibilities and to make adjustments and improvements. Minimum results of these activities are a better understanding of what activities are expected and what goals and objectives are to be met. Needed encouragement may also be communicated. The counselor may find that no amount of training will qualify an incumbent for his position. In that case, the sooner the person is relieved of responsibility, the sooner the organization will be healed.

2. *Transferring* an individual to another position in which his skills and knowledge are sufficient for the duties of the position. If the individual sees the transfer as a promotion or at least as a lateral move, he is far more likely to accept it and to perform well in the new position. A transfer is more effective when the performance of the individual is assessed within a few months.

In a transfer, the individual may be moved from a line to a staff position. Such a move has a greater chance of success if it means that the individual's abilities and talents are used more effectively, even though on an advisory basis. The individual can maintain his dignity, morale, and loyalty if the staff position carries prestige (a high-sounding title) and he suffers no reduction in salary. A transfer from line to staff actually makes the experience and knowledge of the individual more available to line managers when they have special problems. Salaries for staff persons, however, are normally part of overhead and such transfers are economical only when an individual has contributions to make to achieve the overall objectives of the firm.

3. To avoid the sense of failure associated with *dismissal,* the individual may be given reason to resign through unsatisfactory performance evaluations, stripping the individual of meaningful responsibilities, or omitting the individual from conferences and channels of information. These devices are less than honest, they take considerable time to implement, and the individual still may not resign. Furthermore, most of the harm is done to the employees who are putting forth great effort. The individual is harmed because he is prevented from growing and enjoying his work. The manager's reputation suffers because he fails to make a change that is obvious.

Implementing Organizational Changes

To make effective organizational changes, the following steps are involved:

1. Review, or if necessary, determine the objectives the organization is to achieve. Also review the activities of key managers and employees that are required to achieve the objectives.
2. Obtain information on the status of activities, responsibilities, and relationships.
3. Prepare an action plan for organizational renewal that will be necessary to achieve the objectives during a reasonable time.
4. Prepare a schedule of specific times during which each phase will be implemented and establish methods of monitoring those changes.
5. Gain acceptance of the plan by the participation of key people, by extensive communication of the changes with reasons, and by inviting questions, suggestions, and ideas. Modify changes as needed and implement the final plan for change.
6. Changes can frequently be timed effectively to occur with modifications in processes or products.

SUMMARY

Organizing involves defining activities that are necessary to achieve planned goals or objectives, obtaining resources, establishing formal reporting relationships among people, and allocating plans and subgoals to those people. The engineer turned engineer manager finds that he must give subordinates a degree of independence.

Organizing is the first step toward making plans effective. It requires the engineer manager to divide work into meaningful and manageable units and assign each unit to specific groups or individuals. If each unit fulfills its obligation effectively, the time spent with the manager is reduced. Time spent in personal contact is controlled through planning, training, communication, delegation, and evaluation.

A number of methods are available for the assignment of personnel, including grouping by numbers, functions, geographical areas, product, customer, and process. This is the first step necessary toward the preparation of an organization chart. Such a chart attempts to indicate various levels of authority. It may also designate several different types of organizations. Of these various types, the line organization is the most simple. It shows a direct superior–subordinate relationship. Somewhat more complex and difficult to operate is the functional organization in which tasks are assigned by areas of specialty. It leads to difficulty in that the engineer finds himself reporting to several bosses.

The most frequently used type of organization is the line and staff organization. In general, the line is carrying out certain assigned responsibilities to achieve objectives and the staff is available to provide them expert assistance. A variation of this involves functional authority assigned to certain staff departments. These departments can assert their authority over the line organization in their particular area of expertise. Sometimes this functional authority is carried even further, to what might be called specialized authority, where definite responsibilities are assigned to the staff.

Committees are used extensively in all organizations. They may be put together in an effort to have collective knowledge or expertise of a group, in dissemination of information, the sharing of accountability, participation in the decision-making process, or simply to avoid making decisions.

In evaluating committee performance, it is necessary to review its role, the responsibility of its members, the membership size, the chairperson and the cost of the committee's work.

Organizations change, and the engineer manager must be alert to such changes and revise his organizational structure accordingly. This frequently brings about the movement of people because of their inability to perform. Careful analysis of the operation of the organization is necessary in order to effectively make such changes.

IMPORTANT TERMS

Committee:	A group of individuals appointed or elected to meet together for the achievement of some specific goal or goals.
Functional Authority:	Certain staff units may have authority over line units in areas of their particular expertise.
Informal Group Relationships:	Informal relationships that exist for communication and for action.
Line and Staff Organization:	A line organization accountable for the operation of the unit with staff specialists providing expertise and assistance to help the line achieve the organization's objectives.
Line Organization:	Direct superior–subordinate relationship in which direction flows downward and accountability flows upward.

Organizing:	Defining activities that are necessary to achieve planned goals or objectives, obtaining resources, establishing formal reporting relationships among people, and allocating planned subgoals to these people.

FOR DISCUSSION

1. Describe some of the advantages and disadvantages involved in organizing.
2. What is meant by "functional organization?"
3. What is "specialized authority"? To whom does it belong?
4. What factors must be considered if a committee is to be effective? Explain the importance of each factor.
5. Why are staff units often not utilized by line people and what can the engineer manager do to improve the line and staff relationship?
6. The manager of a research and development department is consulting you about improving the organizational structure of his department. Develop an organizational structure. The department consists of several engineers and support persons including draftsmen, secretaries, and a few support technicians.
7. What is the added dimension which makes organizing in the role of an engineer manager much harder than in the role of an engineer?
8. Several factors influence the number and frequency of contacts which the engineer manager must have with subordinates. Name and describe these factors.
9. There are a number of basic techniques of grouping activities with which the engineer manager must be familiar. List and discuss these techniques.
10. Name and describe the three different types of organization.
11. There are a number of reasons why committees are so widely used in organizations. Name and discuss a few of these reasons.
12. Industry is always changing. The effect of changes on people becomes a real problem for the engineer manager. Name some techniques that he may use to correct such problems.
13. What should the engineer manager do if the informal organization conflicts too much with the formal organization?
14. Why should members of the committee be from the same level in the organization?
15. How can organizational changes be made more effectively?

CASE 4-1

Bypassing the Boss

Tim Casey, an avid trout fisherman, has a week's vacation coming and wants to take it during the opening week of the trout-fishing season. Tim learns from Sally, the department's secretary, that two of the other engineers have already requested and received approval from Rex, the manager of R&D, to take vacations during this week. Afraid that Rex would not approve his request, Tim decided to forward his request directly to Tom Johnson, the vice president, who must approve vacation requests after they are approved and forwarded by Rex. Tom is very friendly with Tim and is his best fishing buddy. Tom failed to note that Rex had not initialed the request and approved it. Several weeks passed before Rex found out by accident that Tim had been approved to go on vacation at this time.

This is just one instance in which Rex's people have gone directly to Tom and received permission to do something. This bothers Rex, and to make the situation worse, he overheard a conversation to the effect that if you want approval and quick, don't waste time with Rex but go directly to Tom.

1. What should Rex do considering the relationship between Tim and Tom and the past instances?
2. Is this situation the fault of Tom, Tim, or Rex?
3. What should Rex do if Tom ignores Rex's complaint when approached?
4. How could this problem have been avoided?

CASE 4-2

Company Policy Interpretation

The company's policy on time off from work is stated as follows: . . . for personal emergency an employee is excused during regular working hours. This policy has been interpreted to mean that personal matters that can be taken care of after working hours do not constitute an emergency.

Recently, Roger came to Rex Smith, his supervisor, and asked, "Rex, I would like to come in late tomorrow so that I can register my new car at the motor vehicle agency. There is not time to do it during lunch hour since there is always a long line of applicants. Tomorrow is my day to drive in my carpool so I will have it to go to the motor vehicle office after dropping off my buddies. I might do it after work but the group does not want to wait for me."

"Sure you can, but come to work as soon as you have registered your car," said Rex.

The next morning while Rex was working on some designs for the Mid-Can project, Ted Butts, manager of plant maintenance, stopped by. "You sure put me in a tough spot with two of my employees. They heard about Roger's taking time off to register his new car. I know of no one in the organization accepting an excuse like that as an emergency. Don't you remember the policy?"

1. Was Rex right in letting Roger off? Why?
2. What do you think of Ted's interpretation of the policy? Of Rex's?
3. Where did the initial failure occur?
4. What can be done to avoid this situation and similar ones from arising in the future?

FOR FURTHER READING

ALLEN, L. A., "The Line–Staff Relationship," in *Readings in Management,* 4th ed., eds. M. D. Richards and W. A. Nielander. Cincinnati: South-Western Publishing Company, 1974.

BLAU, PETER M., and RICHARD A. SCHOENHERR, *The Structure of Organizations.* New York: Basic Books, 1971.

BOWER, M., *The Will to Manage.* New York: McGraw-Hill Book Company, 1966.

BROWN, JAMES D., *The Human Nature of Organization.* New York: AMACOM, 1973.

BURACK, ELMER H., *Organization Analysis: Theory and Applications.* Hinsdale, Ill.: The Dryden Press, 1975.

DRUCKER, PETER F., "New Templates for Today's Organizations," *Harvard Business Review* (January–February 1974), 45–53.

FRENCH, W. L., and C. H. BELL, *Organization Development.* Englewood Cliffs, N.J.: Prentice-Hall, Inc., 1973.

HARVEY, EDWARD, "Technology and the Structure of Organizations," *American Sociological Review* (April 1968), 247–59.

JUN, JONG S., and WILLIAM B. STORM, *Tomorrow's Organizations: Challenges and Strategies.* Glenview, Ill.: Scott, Foresman & Company, 1973.

LITTERER, JOSEPH A., *The Analysis of Organization,* 2nd ed. New York: John Wiley & Sons, Inc., 1973.

MELCHER, ARLYN J., *Structure and Process of Organizations: A Systems Approach.* Englewood Cliffs, N.J.: Prentice-Hall, Inc., 1976.

MOTT, PAUL E., *The Characteristics of Effective Organizations.* New York: Harper & Row, Publishers, Inc., 1972.

NEWMAN, DEREK, *Organization Design.* London: Edward Arnold, Ltd., 1973.

NEWMAN, W. H., and J. P. LOGAN, *Strategy, Policy, and Control Management,* 7th ed. Cincinnati: South-Western Publishing Company, 1976.

URIVICH, L. F., "V. A. Graieunas and the Span of Control," *Academy of Management Journal* (June 1974), 349–54.

WEIHRICH, H., and S. N. TINGEY, "Management by Objectives—Does It Apply to Staff?" *Industrial Management* (January–February 1976), 26.

THE TRANSITION
TO SELECTING
AND MANAGING
PROJECTS

Engineer managers are directly involved in selecting and managing projects. The matrix system of management is an effective technique for engineering projects, but many times the arrangement causes confusion for engineers working on the project as well as for the engineer manager. A thorough understanding of the system will assist the engineer manager in obtaining effective results and reducing problems arising from managing projects.

CHAPTER
HEADINGS

- ☐ Ranking Projects by Functional Aspects
- ☐ Ranking Projects by Financial Aspects
- ☐ Each Firm Has Its Own Methods
- ☐ Need for Matrix Management
- ☐ Matrix Management Has Two Meanings
- ☐ Designing a Matrix Organization
- ☐ Establishing a Matrix Management System
- ☐ Making Matrix Management Effective
- ☐ Failures of the Matrix Management System
- ☐ Systems Management
- ☐ Summary

LEARNING
OBJECTIVES

- [ ] Learn methods of ranking projects by functional aspects.
- [ ] Learn methods of ranking projects by financial aspects.
- [ ] Understand system for selecting projects.
- [ ] Gain insight into the importance of matrix management.
- [ ] Learn how to design and establish a matrix organization.
- [ ] Become aware of what can cause failures in matrix organizations.
- [ ] Understand principles of system management.

EXECUTIVE COMMENT

THOMAS B. ROBINSON
Managing Partner
Black & Veatch, Consulting Engineers

Thomas B. Robinson is managing partner of Black & Veatch, Consulting Engineers. A native of Kansas City, Missouri, he joined Black & Veatch as assistant engineer in 1940. He moved through several engineering positions to become a partner in 1946. By 1956 he was assistant managing partner and in 1973 became managing partner. He is also chairman of the board of Black & Veatch International.

The Organizational Structure at Black & Veatch

Black & Veatch has developed an efficient, client-oriented method of project management, which students and practitioners of engineering management properly would call *matrix management*.

Within the firm we call it the *engineering and management system*. The practices used at Black & Veatch are embodied in systematic written methods and procedures. They include (1) control of project by preplanned procedures, (2) design by systems, (3) completion of design prior to construction, and (4) construction with the use of well-defined plans and specifications.

As with matrix management systems used by other organizations, we have both vertical and horizontal forms of management. Project management works vertically in a straight-line, hierarchical line of supervision. The client is at the top of the vertical line, served by the project manager, and through him the project engineer(s) and project staff. The project detail work is carried out by the various departments represented at the project staff level—the horizontal part of matrix management.

We find that matrix management improves project efficiency, reduces waste time, promotes better and more consistent communications with the client and project team, and results in a better and more cost-effective service to the client.

The matrix system reduces the possibility of a totally new project team reinventing the wheel with each project. Both the vertical/project management part of the matrix and the horizontal/department parts of the matrix make better use of prior experience and knowledge. This is particularly evident on repeat commissions, which represent a high percentage of our work.

We have found that matrix management improves our ability to plan, schedule, and complete our project tasks. That ability means we are better qualified to retain existing clients and attract new clients. The matrix is internal, but its most valuable results are external. It serves our projects, which in turn serve our clients.

Whatever the means or measures selected to determine which among several projects is the most attractive, these measures are attempts to achieve objectivity. Unfortunately, that goal, like the Holy Grail of the Knights of the Round Table, is never achieved. Failure to achieve objectivity, however, does not lessen the engineer manager's need for it. On the contrary, he is in a unique position with the organization. He is aware of the state of the technical art, and with a little study, can be sufficiently aware of the art of economics to evaluate alternative projects, whatever degree of objectivity his organization may require.

Karger and Murdick reproduce several ranking charts devised by chemical and pharmaceutical firms in the United States.[1] One of the charts is shown in Figure 5–1. These firms have found it convenient to rank projects according to their functional aspects, which include financial, production, engineering, research and development, and marketing. Each aspect is broken down into more finely tuned elements of the function. The project is then given a rating from –2 to +2 according to that breakdown. The result is that a marketing aspect such as "effect on present projects" will have a –2 rating if the new product will replace an existing one and a +2 rating if it will increase the sales of other products without eliminating any of those now sold by the company. The pharmaceutical firm to whose rating methods Karger refers uses a numerical scale of from 1 to 3. Whatever the scale, however, the results are the same. The engineer manager, in concert with other interested executives, compares the result of the proposed new

RANKING PROJECTS BY FUNCTIONAL ASPECTS

[1]Delmar W. Karger and Robert G. Murdick, *Managing Engineering and Research* (New York: Industrial Press, 1963), pp. 193–221.

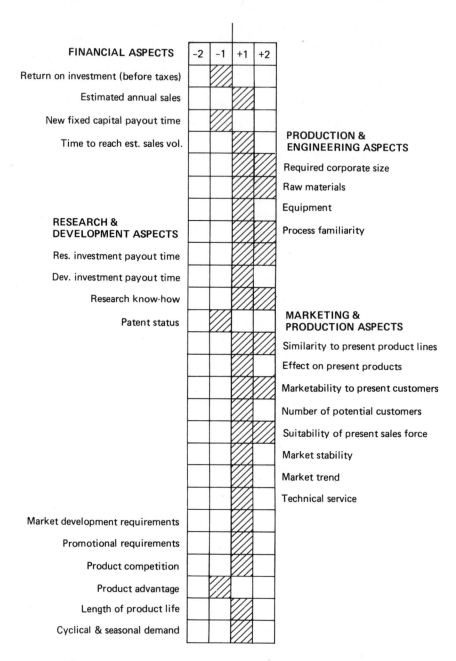

FIG. 5-1. Selecting Projects and Establishing Objectives (Source: John S. Harris, "Selecting Projects." Reprinted, with permission of the copyright owner, The American Chemical Society from the April 17, 1961 issue of *Chemical and Engineering News,* Vol. 39, No. 16, p. 110.)

project to the rating scale and with some algebraic adding can come up with a series of subtotals, one for each of the corporate functions, and with a grand total.

The potential for differences of opinion among the concerned executives is obvious. If the executive representing a particular function were to take either a like or a dislike to a given project proposal, he could affect its numerical rating by his estimate of its relative position for those aspects characteristic of his function. Another potential weakness of the ranking method is that a company's decision to pursue or abandon a given project can seldom be made on the basis of what will happen within the next year of two. On the contrary, those decisions will affect market share, sales volume, and profits for many years, maybe even for decades. It is that future aspect of particularly high-technology products and processes which makes a ranking method so subject to error. Even the most skilled of market development and product research executives may not be able to combine their skills so as to foresee the entry of a given material into completely new markets beyond those considered at the time of the ranking. Engineer managers may not be able to foresee scientific breakthroughs which might be applied to a new product with minimum extra costs and thereby reduce its selling price or expand its sales into new and unforeseen markets.

Importance of Ranking

The engineer manager may at this point ask, "Is it worth our time to go through a ranking?" The answer is yes, not because of the validity of any set of subtotals, but because of the chance to communicate with other executives in the organization about the possible future of a new venture long before the firm is committed financially, technically, and often emotionally to its development. Further, it gives the engineer manager an opportunity to identify technological gaps in processes, materials of construction, and application methods that affect the success of a new product. A manager can then be alert to the filling of one of those gaps by research, either within or without the company. For example, he can give more intelligent counsel to his management relative to the acquisition of a license to manufacture or use a product that will materially assist in the introduction of another product that his company has been considering. The plastics industry is full of examples of such developments which, though made by one company, have been licensed or used by another to further its own marketing end. For example, for years fiberglass reinforced skylight panels were notorious for their lack of resistance to ultraviolet radiation from the sun. It was not until the DuPont Company developed its "Tedlar" polyvinyl fluoride film and until the Butler Manufacturing Company learned how to apply that film to the skylights during their manufacture that the plastic skylight industry was able to escape its problems with sunlight-accelerated weathering.

Although rating scales are used with new product profiles, and although these scales do generate numbers, the project rating profiles are more accurately describable as more qualitative in their rankings than quantitative. Stated in the terms that introduced this section, ranking projects by functional aspect profiles may be so far from objectivity that the Holy Grail is not even visible.

RANKING PROJECTS BY FINANCIAL ASPECTS   As an adjunct to or alternative to ranking projects by functional aspects, the engineer manager can rely on one or more of several purely economic yardsticks. Sometimes those yardsticks are so simple that the primary technical and economic tasks are those associated with finding the least expensive equipment that is adequate to replace something that has broken down. Similar in their simplicity are those situations in which a corporation must invest to meet the regulations of governmental agencies, such as the ones concerned with the cleanliness of the environment. Once again, the question is not one of ranking among projects, but of among pieces of equipment which are technically sound but vary in their cost, availability, probable maintenance requirements, and supply of parts. A fire or a natural disaster may leave a company little choice among investment alternatives because the primary consideration is that of minimizing the time to get back into business. Less apparent to the engineer manager is the need for projects that will remedy a competitive weakness of a company. In such instances, he usually relies on counsel from marketing executives to assert the degree of need for a given project.

Selection by Payout Time

Park describes in detail a method of selecting the most attractive project based on payout time (Tables 5–1 and 5–2 and Figure 5–2).[2] Under this method, a company may calculate a cumulative cash flow (income minus expenditures), may discount these values at some rate associated with the current time value of money, and may thereby determine how many years it will take before the company has earned its money back. At this point, the engineer manager will be able to use these techniques more effectively if he is familiar with the methods of depreciation which are allowed by the Internal Revenue Service and favored by his company.

In addition to ranking projects by their payout time with or without discounting the cash flow, most companies that use the criterion of payout time also have a maximum number of years they will allow. Normally, that number is three. Stated otherwise, if they cannot see that a project will return its investment within three years, they are not interested in it even though it may be the best of a half-dozen alternatives currently available to them.

Selection by Rate of Return

A more involved method of financial comparison of projects is that of the rate of return. Park describes an accounting approach in which the profit after depreciation and taxes on a project is related to its total or average outstanding

[2]William R. Park, *Cost Engineering Analysis* (New York: John Wiley, 1973), pp. 31–37.

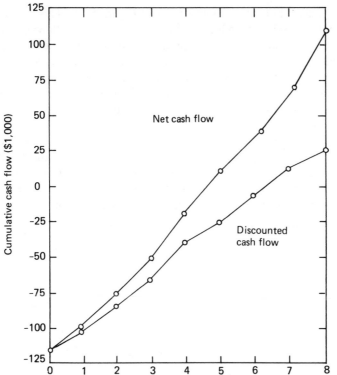

FIG. 5-2.   Cumulative Cash Flow for Project A. [Source: William R. Park, *Cost Engineering Analysis* (New York: John Wiley & Sons, 1973), p. 35.]

TABLE 5-1
Financial Summary of Proposed Project

Year	Cash Outlay	Net Income Before Taxes and Depreciation	Depreciation Charges	Net Taxable Income	Income Taxes	Net Income After Taxes	Net Cash Flow
0	(120,000)	0	0	0	0	0	(120,000)
1	0	20,000	15,000	5,000	2,500	2,500	17,500
2	0	30,000	15,000	15,000	7,500	7,500	22,500
3	0	40,000	15,000	25,000	12,500	12,500	27,500
4	0	50,000	15,000	35,000	17,500	17,500	32,500
5	0	50,000	15,000	35,000	17,500	17,500	32,500
6	0	50,000	15,000	35,000	17,500	17,500	32,500
7	0	50,000	15,000	35,000	17,500	17,500	32,500
8	0	50,000	15,000	35,000	17,500	17,500	32,500
	(120,000)	340,000	120,000	220,000	110,000	110,000	230,000
							−120,000
							110,000

Source: William R. Park, *Cost Engineering Analysis* (New York: John Wiley, 1973), p. 33.

investment period (Tables 5–3 and 5–4 and Figure 5–3).[3] This method does not use discounted cash flow, which means that it makes no allowance for the time value of money. Another variation on the rate of return concept is the operating return, which expresses the rate as the ratio of the average cash return to the original investment. This method is particularly good for comparing the economic operating effectiveness of the different parts of a business. It measures each part's cash contribution to the business, but it does not show the economic value of a given project.

There is a method of calculating the present worth of a company if it successfully undertakes a project versus the successful undertaking of another project, and even versus the undertaking of no new project. The present worth method determines the value now of cash flows in the future that may occur because of a given project. A special case of the present worth method is one in

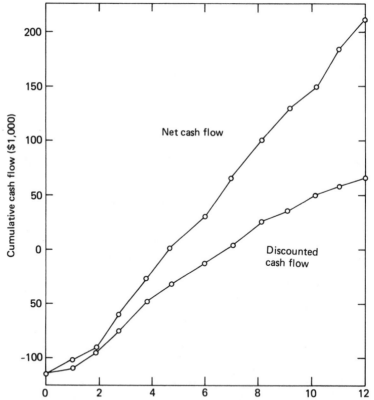

FIG. 5–3. Cumulative Cash Flow for Project B. [Source: William R. Park, *Cost Engineering Analysis* (New York: John Wiley & Sons, 1973), p. 35.]

[3]Park, *Cost Engineering Analysis*, pp. 38–39.

TABLE 5-2

Cash Flow Summary (Project A)

Year	Net Cash Flow	Cumulative Cash Flow	Net Cash Flow Discounted at 10%	Cumulative DCF
0	(120,000)	(120,000)	(120,000)	(120,000)
1	17,500	(102,500)	15,900	(104,100)
2	22,500	(80,000)	18,600	(85,500)
3	27,500	(52,500)	20,600	(64,900)
4	32,500	(20,000)	22,200	(42,700)
5	32,500	12,500	20,200	(22,500)
6	32,500	45,000	18,300	(4,200)
7	32,500	77,500	16,700	12,500
8	32,500	110,000	15,200	27,700

Source: Park, *Cost Engineering Analysis*, p. 34

TABLE 5-3

Financial Summary of Project B

Year	Cash Outlay	Net Income Before Taxes and Depreciation	Depreciation Charges	Net Taxable Income	Income Taxes	Net Income After Taxes	Net Cash Flow
0	(120,000)	0	0	0	0	0	(120,000)
1	0	20,000	10,000	10,000	5,000	5,000	15,000
2	0	30,000	10,000	20,000	10,000	10,000	20,000
3	0	40,000	10,000	30,000	15,000	15,000	25,000
4	0	50,000	10,000	40,000	20,000	20,000	30,000
5	0	50,000	10,000	40,000	20,000	20,000	30,000
6	0	50,000	10,000	40,000	20,000	20,000	30,000
7	0	50,000	10,000	40,000	20,000	20,000	30,000
8	0	50,000	10,000	40,000	20,000	20,000	30,000
9	0	50,000	10,000	40,000	20,000	20,000	30,000
10	0	50,000	10,000	40,000	20,000	20,000	30,000
11	0	50,000	10,000	40,000	20,000	20,000	30,000
12	0	50,000	10,000	40,000	20,000	20,000	30,000
	(120,000)	540,000	120,000	420,000	210,000	210,000	330,000
							-120,000
							210,000

Source: Park, *Cost Engineering Analysis*, p. 36.

TABLE 5-4
Cash Flow Summary (Project B)

Year	Net Cash Flow	Cumulative Cash Flow	Net Cash Flow Discounted at 10%	Cumulative DCF
0	(120,000)	(120,000)	(120,000)	(120,000)
1	15,000	(105,000)	13,600	(106,400)
2	20,000	(85,000)	16,500	(89,900)
3	25,000	(60,000)	18,800	(71,100)
4	30,000	(30,000)	20,500	(50,600)
5	30,000	0	18,700	(31,900)
6	30,000	30,000	16,900	(15,000)
7	30,000	60,000	15,400	400
8	30,000	90,000	14,000	14,400
9	30,000	120,000	12,700	27,100
10	30,000	150,000	11,600	38,700
11	30,000	180,000	10,500	49,200
12	30,000	210,000	9,600	58,800

Source: Park, *Cost Engineering Analysis*, p. 37.

which a discount rate is determined which when applied to the cash flows makes their time values total zero. This method determines a return on investment. That return may be called the internal rate of return, the profitability index, or the discounted cash flow return. Here, as with the maximum payback period, the company may set a minimum rate of return below which a project will not be considered no matter how attractive it is for other reasons. There are some technical disadvantages in using the internal rate of return method, but they have not been sufficient to discourage its application completely.

Possibly the closest approach to complete objectivity in the economic comparison of engineering projects is that offered by calculating the return on investment. Park presents a thorough description and critical analysis of the method.[4] He also points out that there are computer programs to avoid the trial-and-error method originally used in calculating it and that there are some graphical shortcuts to approximate the return on investment of a series of projects. As with the payback period, companies will usually set a minimum for the return on investment below which they will not accept the project. Further, a few corporate management groups have learned to insist on an analysis of return on investment calculations according to the sensitivity of their results. That sensitivity relates to the effects of the unexpected. Return on investment is affected by variations in working capital, fixed investment, variable manufacturing costs, sales volume, equipment costs, and the time required to get into production and for the product to reach a given level of sales. All of these factors involve uncertainty. For this reason, the engineer manager will deliberately vary estimates of the components of the calculations and may even assign probabilities to those estimates. The re-

[4]Park, *Cost Engineering Analysis*, pp. 40–45.

sults of his variations can sometimes draw him even closer to objectivity by demonstrating areas in which hastily drawn estimates may affect the ranking of a given project versus that of others.

No matter what financial ranking method is preferred by the engineer manager, his company will have its own preference of which one to use and sometimes of which part of the company shall do the analysis. It will behoove the manager to learn those ground rules and live according to them. It will also behoove

EACH FIRM HAS ITS OWN METHODS

him to keep at least some caveats in mind. One is that one or more parts of an organization may develop an antipathy or a liking for a project to the point that either position can affect their objectivity in contributing their estimates to the calculation of return on investment. Sales volumes can be modified upward relying only on the enthusiasm of the marketing arm for the potential product. Conversely, the time to get into production measured from the time when expenses and investments are first charged against the project may be stated so conservatively that the net cash flow of a project will become positive so many years in the future that its return on investment becomes unattractive. The second caveat is an exaggerated case of the first. It, however, is usually found in a positive vein and occurs when one executive in a company becomes so enamored with an idea or a new process that he becomes its "champion." Objectivity becomes less important to him than persuading other executives within the firm to accept his evaluation of the proposed project. The engineer manager may be well advised, in fact, not to proceed very far with any project unless it *does* have a champion who is high enough in an operating division to make the investment and expense decisions that are necessary for the success of the project. Looked at in a different way, the engineer manager may find himself seeking a project champion before presenting it on any broad scale. The risks are obvious. The Holy Grail having been replaced by emotion, is farther away than ever. The emotion is not all bad; on the contrary, successful products on the market today would never have passed any of the screening tests previously mentioned had they not benefited from the persistence of someone who believed in them. They would have remained in the recommendation section of a progress report and would never have seen the light of commercial day.

To handle large engineering projects effectively, it is frequently necessary to group engineers from several different departments or specialties under a project leader or manager for a limited period of time. Such groupings are most effective if the firm itself is organized as a matrix of functional and project

NEED FOR MATRIX MANAGEMENT

management. In our discussion of matrix management the term "engineer manager" applies either to the project leader or the department head.

When the engineer manager was an engineer, he often had a dual role. That duality was his introduction to matrix management. Not only did he function as a member of his department, he was often a member of one or more project teams. The projects may have varied from very small to large complex ones. Depending on the magnitude of the project, the engineer was either a member of a large project team reporting to a project manager or was a member of a small group with no identifiable project leader or project organization. Whatever the size of the project team, the engineer filled a dual role. He represented his department on the team and was accountable both to his department and the project manager. As a member of the team, the engineer performed well-defined duties or tasks. He was required to keep an account of the time he spent. Sometimes he was assigned to more than one project with a commensurate increase in the complexity of accounting for his time. The dual role existed only for the duration of the project. After its completion, he was again accountable only to his department manager.

As a member of the project team, the engineer was responsible for meeting the objectives set forth by the project manager. In this role, the engineer was able to define or at least identify his special tasks in the project and was directly involved in helping the project team meet its objectives. As a team member, he depended on others and others depended on him to achieve those objectives. No single engineer could meet the project objectives by himself.

If he was called on to perform the engineer manager functions of a project leader, he was no longer concerned only with specific tasks or parts of the project. Instead, his concern became the overall project and the performance of all members of the project team. He now had to plan the project, organize the available resources, and coordinate the activities of the team members. Frequently, he found himself doing part of the engineering as well as acting as an engineer manager.

The project manager usually carries the primary responsibility for accomplishing the goals of projects. As industry becomes more technically oriented, its projects require expertise from many highly specialized areas. The job of the project manager becomes thereby even more important but coincidentally more difficult. That increase in difficulty makes it more important than ever for the engineer manager to thoroughly understand project management in a matrix organization. The matrix aspect means that the project manager will direct teams of highly specialized members of his own and other departments. It also means that the members of his team may be involved in various projects in other departments.

MATRIX MANAGEMENT HAS TWO MEANINGS

All matrix organizations have two sets of executives. More specifically, the term has one meaning in which one set of executives directs the day-to-day operations of product or service divisions and another meaning in which the other set directs certain longer-range aspects of such functions as manufacturing engineering, and marketing. Within that kind of matrix organization there are project teams, but they are either so short-lived or unimportant that they escape recognition in the organization chart.

It is the second meaning of the term that is of concern here. In such organizations the success of projects is vital to the success of the firm. There are still the two sets of executives. Within an operating division, however, there is now a set of project managers. The matrix of primary importance now becomes that involving project managers on one axis and department managers on the other. Such a matrix may exist entirely within the engineering department of a large manufacturing company, as shown in Figure 5–4.

The matrix organization finds its greatest use in managing complex activities requiring the planning, organizing, staffing, controlling, and directing of engineers of differing specialties and experience. Figure 5–5 shows a more generalized matrix type of organization.

In a manufacturing firm, the horizontal row of manager boxes will be filled with the names of department or functional managers. The vertical manager boxes will be project managers. Like the engineers, the project managers are often in a dual role in that their organizational "home" may be in one of the functional departments.

In a consulting architectural/engineering firm, the top row of manager boxes will usually be project managers. Although not shown in Figure 5–5, many firms will use project engineers as intermediate figures between the project

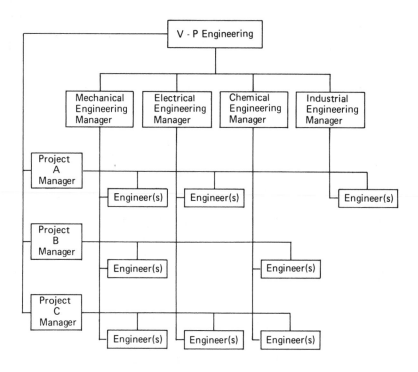

FIG. 5–4. A Project-Discipline Matrix Within an Engineering Department

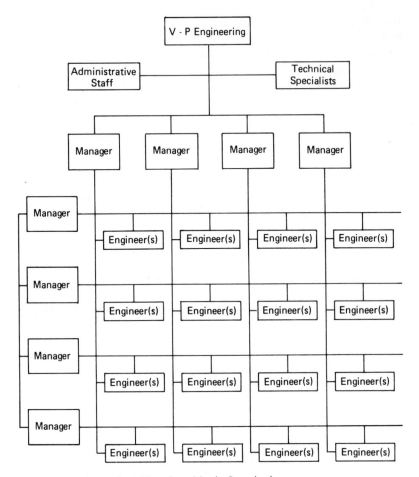

FIG. 5-5. A Generalized Chart for a Matrix Organization

manager and the engineers. The vertical row of manager boxes will be for the heads of engineering departments (disciplines).

Thomas B. Robinson, Managing Partner, Black & Veatch, offers the following comments on matrix organization as practiced by a consulting engineering firm:

> For years at Black & Veatch we had strictly vertical project management. Committing engineers to projects was a simpler task when our firm was smaller, not just because of organization size but also because at that time we could handle work solely on a project basis.
>
> When we were selected by a client, we reassigned engineers to staff that client's project. We had a project manager, a project engineer, and an appropriate number of study or design engineers. That degree of organization sufficed to complete the projects, to satisfy our clients, and to encourage our clients to retain us again when they next needed engineers.

But with that success came growth in the number, size, and complexity of projects and the need for more specialists, nontechnical, and support personnel. Today, only one of our divisions is still small enough to function effectively on the traditional project line management system.

Project work in the other larger divisions has evolved into matrix management. The executive partner in charge of each division has two sets of subordinate executives: project managers and department heads.

The matrix also means that our individual engineers have two bosses: one is the project manager or his project engineer(s) and the other is the head of a department, which is staffed either with engineers of a common discipline, such as mechanical, or of a common project responsibility, such as systems.

Let's examine how the matrix system works in our largest division, the Power Division. The executive partner and head of the Power Division assigns a project manager who reports to him and has complete responsibility for the project. The executive partner and the project manager then assign project engineers to the project.

Although working within the various engineering departments, each project engineer is responsible to the project manager for a particular task. Other engineers within a department may be assigned to the project engineer depending on the complexity of the design. This organizational structure has the advantage of retaining the benefits of departmental functions (i.e., personnel administration, training, and technical discipline) while maintaining the direct line responsibilities of a project group.

In addition to the project manager, project engineers, and staff, each project team includes assigned systems engineers from the Systems Engineering Department and an assigned project lead design engineer from each department.

The respective departments provide engineering, drafting, shop drawing checking, and other support services as required. The advantages of performing detailed design and technical production work in organized departments are that the departments provide flexible manpower resources, provide accessible centralized files, and make possible higher levels of specialization and supervision than if the various disciplines are each fragmented into several project groups.

We at Black & Veatch have had enough experience with matrix management to recognize its flaws as well as its advantages. It places heavy responsibility for communication and coordination at two major points: the project manager/project engineers and the department heads.

The matrix appears to violate a tradition of good management in that the individual engineer has two bosses instead of one. This has not been a problem in our organization because of the measures described earlier. The reverse side of having two bosses is that it exposes the good work and possible promotion of an individual to two supervisors rather than one.

Characteristics of Project Management

Whatever the nature of the enterprise or project, the important characteristics of project management are as follows:

1. The goals and tasks for the project must be defined in terms meaningful to both engineers and project managers.
2. The project has a limited life.

3. The project requires more frequent progress reports than do routine departmental activities.
4. The project manager requires information on costs and on mandays expended.
5. The project manager requires correlations of expenditures with task progress data.
6. The goals of the project change frequently.

Most of the above characteristics could be achieved more effectively if there were only project teams and no matrix organization. The elimination of the functional managers would mean lower fixed overhead and much less time spent in coordination and communication. Accountability (credit or blame) would be much easier to assign by the chief executive. A "pure project" organization, however, is not feasible in most firms because (1) highly specialized technical personnel cannot be fully utilized and (2) expensive but useful equipment may not be justifiable under any one project.

Attractiveness of Matrix Management

On balance, the matrix organization is more attractive than a pure project organization because:

1. It provides "pools" of technical expertise which lets an organization concentrate its resources on an area of need.
2. Day-to-day activities of the organization are maintained.
3. The project team members have an organizational "home" to which they return if no other project assignment is available. For that reason, their morale and efficiency do not suffer near the end of a project.

Advantages of Project Management

Whether the firm uses matrix or pure project organization, emphasis on project management has a number of major advantages:

1. There is closer control of resource allocation and consumption.
2. Closer identification of progress with use of resources is provided.
3. Responsibilities and accountabilities are more precisely defined.
4. The relationship of individual efforts to task accomplishment is clearer.
5. There is better quality control.
6. Goals are accomplished in the shortest possible time.

7. Innovative individuals can be identified.
8. Continuity is created in the pursuit of project goals but with allowance for personnel job changes.

Disadvantages of Project Management

Heavy emphasis on project management includes such major disadvantages as:

1. Qualified project managers are rare and expensive.
2. A matrix organization creates complex relationships which are not easily understood by most engineers.
3. Project managers may resist termination of their project and transfer of personnel and resources to other teams.
4. Delay in any phase may idle expensive engineering talent.
5. The constant pressure to get things done on time can adversely affect the health and performance of project leaders and team members.
6. The lack of easy communication among project managers can lead to duplication.
7. Higher salary costs are incurred to carry two classifications of engineer managers.
8. Engineers with more than one boss may have no boss.

Project management organization is a response to the problems of applying high technology. The matrix organization is a response to the needs for status and authority for project managers. All engineer managers need to be able to design or at least to understand the design of a matrix organization. The difficult **DESIGNING A MATRIX ORGANIZATION** part of such understanding arises when the engineer manager considers alternatives for grouping activities. Chapter 3 points out that activities can be grouped under headings such as product, marketing, engineering, common processes, geographic, etc. Each of these has management, cost, and economic advantages and disadvantages associated with it.

Whatever the grouping, the project capability of a matrix organization provides quick response to problems or opportunities related to a particular product line or service. Such responses require coordination among specialized groups to achieve project objectives. The functional capability of a matrix organization facilitates the maximum utilization of specialized resources by making them available for several projects. In general, fewer resources are necessary than if the matrix did not exist.

In designing a matrix organization, the effective engineer manager keeps in mind that both performance and coordination of efforts are important. Those advantages are not always easily obtained. One reason for this difficulty is that the engineer manager may have been accustomed to a functional organization. The independence of the function often means that the coordination of the flow of work through it is erratic. The engineer manager can rise above such a background by informally integrating the functional specialties to respond to changes in new technology and to meet schedules. But as operations become more complex and technical, exclusively functional organizations break down. To overcome these problems, the engineer manager forms a *task group* with representatives from all the major departments to enter into joint decisions. The task group is intended as a vehicle for making decisions at the lowest possible level by the most knowledgeable people. Ideally, therefore, the task group is made up of individuals with the authority and information to make decisions and solve problems. Further, the task group must be given time away from other responsibilities to devote full time to solving problems and working with the group. It is but a short step from task groups to permanent teams.

The Team Approach

A team approach may be initiated by the engineer manager around either functional or product lines. These teams are permanent and meet on a continuous basis to solve problems that are of interest to more than one function. In addition to the teams, task groups are still used to solve temporary problems. The teams may consist of engineer managers from different levels in the hierarchy with corresponding discretionary limits to their authority. This multiplicity of levels allows a greater number of day-to-day operating problems to be solved by lower echelons.

As technology and markets change more rapidly, the greater the necessity for flexible, responsible strategies and for faster decision making. In today's fast-changing world these needs are common occurrences in the life of the engineer manager. In the functional organization structure, the need for flexibility and speed is more extensive and is coupled with the need for specialized attention to joint decision making and shared responsibility. Therefore, the engineer manager tries to ensure that all of the concerned functional units are considered in the decisions so that coordination is achieved.

The team concept is intended to achieve coordination of various activities and departments. At times, however, teams can be rendered ineffective by conflicts among team members. The effects of such conflicts are compounded by their dependence on technical competence rather than formal authority to be effective. The seemingly difficult problems associated with lack of formal authority are met with the development of a matrix organization. The authority of the project manager facilitates solving these problems that require action from engineering, production, and other functional units.

Project managers have influence and status similar to those of functional managers. The importance of the project manager can be found in greater participation in budget planning and later approval of the budget and in the development of staff with expertise in accomplishing the goals and objectives.

To the engineer manager the matrix organization provides coordination among specialized groups and more effective decision making, yet still permits the necessary maximum specialization in an organization having new projects, innovations, and constant changes. The matrix organization provides the project manager a dual relationship and balance between the functional areas and the project areas. Thereby, the project manager has input into budgets, resource allocations, salaries, and information systems. Also, he is able to make relevant decisions and achieve coordination, decisions are made at lower levels, and input from functional areas make them joint decisions.

Need for Project Management

A matrix organization can take on many forms which are determined by the following characteristics:

1. *Diversity* of projects undertaken by the organization. The greater the diversity, the greater the need for project management. With diversity, it is difficult for the functional manager to maintain expertise in all technical areas. Decisions on resource allocations, schedules, priorities, and trade-offs all demand immediate attention and knowledge in specialized areas.

2. Degree of *specialization* in functional areas. The higher the degree of specialization, the greater the need for a project arm in the organization. The reason for that need is twofold: specialization breeds isolation and one unit may have a direct impact on the accomplishment of another unit's goals.

3. Changing *competitive conditions.* Such changes frequently require decisions on project goals and organization which may affect several specialized areas.

4. Degree of *technology* utilized and rate at which new technology is being introduced in the industry. To coordinate activities with high levels of technology effectively requires a project manager.

5. *Size* of operation. The large operation may take advantage of economy of scale, highly specialized equipment, and production facilities, but its functional organizational form may make the project manager a necessity to achieve coordination of all activities.

The purpose of matrix design is the optimum allocation of resources to achieve objectives while remaining within the defined constraints. The design involves developing the major tasks, resources, organization, and control systems. The project design is a detailed plan for implementing the project. Its development includes the following activities:

    1. Defining specific technical, cost and personal objectives of the project.
    2. Defining personnel utilization for the various functional skills required.
    3. Scheduling materials and facilities to avoid critical delays.

4. Budgeting and cost estimating so the group knows when it is time to take corrective action, e.g., before costs are out of control.
5. Detailing time schedules that show the time required to accomplish each major task. These time schedules must be closely related to budget and material resources.
6. Measuring performance and variations from design plans. The objective is to recognize trouble areas early enough that action can be taken to avert serious problems.

ESTABLISHING A MATRIX MANAGEMENT SYSTEM Project teams are established to devote special attention to developing a new product or to correct product designs prior to initiating construction projects. Therefore, the objective is to give special engineering management attention and emphasis to projects or problems. The project will generally require people and resources from existing company units. The project management approach can be effectively utilized for one-time activities that:

1. Have a very specific task or goal
2. Involve complex projects that require expertise from several different areas
3. Must be accomplished successfully as a matter of major importance to the company
4. Are not routine

The matrix organization is developed to avoid weaknesses inherent in an exclusively functional organization, which may be:

1. Concerned only with accomplishing specialized jobs within a given budget
2. Concerned only with its special expertise rather than project activities.
3. Concerned only with its own portion of activity
4. Not concerned with the project's cost and the resulting profits
5. Unprepared to be flexible and responsive to rapidly changing project requirements
6. Unaccustomed to making the quick decisions required for the project

Approach to Establishing a Matrix Organization

The establishment of a matrix organization must be approached in an orderly manner to avoid many of the unpredictable events that inevitably occur. Also, some problems can be foreseen on the basis of past experience, and alternative approaches can be developed, thereby avoiding many crises. Steps in establishing a matrix organization are to:

1. Assess the nature of the problems and opportunities facing the organization
2. Evaluate existing organizational weaknesses relative to project implementation
3. Develop a matrix organization to overcome these weaknesses
4. Determine the matrix organization according to the work to be accomplished
5. Select a project manager who has superior:
   A. Technical experience
   B. Skills in planning, budgeting, and scheduling
   C. Leadership ability
6. Select a project manager who already has a high position of responsibility or reports to the same level as do the department managers from which the project team members will be drawn
7. Provide a project manager who has adequate time to devote to the project
8. Establish adequate cost controls
9. Give the project manager proper support when he deals with the functional units
10. Reward employees involved in projects, for example, by
    A. Salary increases
    B. Promotional opportunities
    C. Relocation of personnel phasing out of existing projects in a manner satisfactory to them

For the matrix organization to be effective, the project manager must have the necessary authority to:

1. Direct the resources to accomplish the objectives of the project
2. Obtain resources for the project
3. Allocate funds and personally perform tasks of the project
4. Schedule and coordinate the project requirements involving the company, its subcontractors, and its clients
5. Solve problems relating to the project and recognize significant problems
6. Develop project controls and monitor progress in relation to time, cost, and degree of completion
7. Approve all project changes
8. Coordinate all subcontractors involved in the project
9. Coordinate all outside contacts related to the project
10. Utilize, if feasible, new technology affecting the project

All of these are essential for the project manager to be fully accountable for the project's success.

The matrix organization may be very formal and have written job descriptions and responsibilities. In most cases, the project manager must rely on a project team staff provided by the functional departments. This staff reports directly to the project manager during the project. The project manager's degree of success depends on his relationship to the functional departments involved (see Figure 5-6).

No definite rules can be set forth for determining the project manager's degree of authority. In some cases, the project manager is able to set schedules, define performance criteria, and control funds, and thereby he can adequately control a project even though no one involved reports directly to him. In other cases, the engineers and support personnel are assigned to the project manager for direct supervision in matters specifically related to the project.

In general, functional units dislike the matrix concept because it superimposes the project organization on their operating structure. Moreover, the functional departments are aware that experience and know-how reside in their personnel. In a matrix organization, the functional departments feel that they become service units and thus give up some of their authority. Adjustments for their personnel and managers are difficult.

The degree of authority given the project manager depends on the project's organizational requirements. Top management must clearly define and communicate the limits of the project manager's authority and his relationships with functional units. In addition, top management must give the project manager the support necessary in his exercise of this authority. If top management does not do this, the project manager and functional unit managers will be constantly engaged in a battle to determine who runs the project.

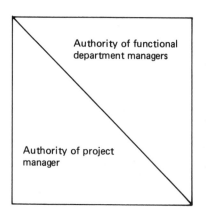

FIG. 5-6. Authority of Project Manager Versus That of Functional Department Manager

Many individuals not working in engineering areas fail to appreciate the value and versatility of this form of organization. They do not recognize the need for matrix organization to deal with specific problems. Therefore, few companies have learned to use it effectively and thereby enjoy the benefits. These benefits include solving problems faster and completing projects and major expansions on schedule and within projected costs.

Matrix management is not a cure-all; to be effective, it must be properly managed by all concerned, including top management. While there are no specific rules, management can insist on answers to certain questions to ensure the success of such an undertaking. They can ask:

1. Is there a need for matrix management? An undertaking is feasible if:
   A. It is a one-time situation with a single and specific end point
   B. It is a complex undertaking not suitable for handling within the present organization

2. Is the project dependent on several functional units? Its completion requires personnel and resources from several units which affect the cost of each task.

3. Is the project critical? That is, will the failure to complete the project on schedule or within costs cause serious losses or penalties?

4. Is there an individual selected who can be responsible for the planning, coordination, and outcome of the project? An engineer manager who tries to manage one of his own projects must devote more of his time than he can generally afford. At the same time, he will not be providing the project with the concentrated attention it demands.

5. Is there adequate interdependence and flow of lateral information among functional units? The project must be characterized by strong lateral working relationships.

If these five items are seriously considered in establishing the project, a significant step is made toward ensuring a successful matrix undertaking. They are necessary conditions that must be considered when developing the organization rather than afterward. Such planning will ensure the full benefits of a matrix management organization.

Matrix management cuts across organizational lines and, in a sense, conflicts with normal organization concepts. It requires the project manager to develop working relationships among departments at all levels. To ensure successful completion of projects, it is possible to set forth certain matrix concepts. These may be summarized as follows:

1. Develop meaningful working objectives of the project to include:
   A. The company's intent for the undertaking and how that undertaking will complement or help achieve other goals

B. The resources and departments involved in the project

C. A description of what the project is expected to achieve or accomplish

2. Develop an organizational structure that reflects the project's objective.

3. Make proper assignments in light of responsibilities.

4. Require reports that use a consistent format with which management is familiar. These reports should include time, costs, and other pertinent information in an easily understood style. These reports may involve a precedent network for scheduling, such as CPM or PERT.

5. Develop a communication network system to keep all involved informed of the project and its progress.

Crystallizing these concepts will make it far easier for the project manager to successfully complete the project and to develop effective working relationships with the functional departments. However, managing a fast-moving project can be difficult for even the most experienced project manager. He must keep track of what is happening and play a crucial role as advocate of the project. Therefore, it is essential that top management create an environment that will allow the project manager to be effective. The project manager can then assure rapid progress under pressure and constant changes, yet be able to keep top management informed of the project's current status.

FAILURES OF THE MATRIX MANAGEMENT SYSTEM

The matrix management system may cause difficult and unusual problems for the project manager in trying to coordinate the contributions of the various groups involved in the project. It is difficult to list all of the problems that may exist in such systems, but some of the most common ones are summarized below for the benefit of both project managers and department managers:

1. Failure of management to develop an organization to meet the specific requirements
2. Failure of management to define the responsibilities, assignments, and relationships among the groups necessary to achieve the goals
3. Failure of management to define working relationships with departments
4. Failure of top management to give adequate support to make the project a success
5. Failure of top management to realize the effect of ill-advised intervention
6. Failure of top management to be realistic when developing the system and assigning responsibility to the project manager
7. Failure to assign the best possible person as project manager

8. Failure of top management to give the project manager's position appropriate status
9. Failure to develop a system that will give prompt information to all involved
10. Failure of project management to develop plans and to determine optional alternatives
11. Failure of top management to give authority to the project manager to cope with contingencies quickly and effectively.

As can be seen, the causes of failure can be traced in large part to top management's not really taking the time and effort to properly establish the system and to give it adequate attention during its lifetime.

In recent years the technical manager has been faced with the problem of managing much larger and more complex technical projects. As the complexity has increased, a larger number of people with widely different backgrounds and skills have had to concentrate their efforts on the project. This has led to problems of communication, coordination, and integration of efforts.

SYSTEMS MANAGEMENT

A system involves a set of elements united in some manner to perform a set of designated functions in order to achieve desired results and it represents the way these complex projects must be handled.

Systems management or the systems approach to management has been developed to meet the needs of these complex projects. Systems management deals with the totality of everything required to bring about the desired result. It deals with the interrelationship of all of the work elements (subsystems) involved.

The systems approach to the management of complex projects recognizes the interrelationships that tie a system together in the environment in which it must operate. It considers the impact of physical, social, political, economic, and technological factors. These environments provide certain constraints on the use of the system and of the technology that must be considered in the design, development, and operation of the desired system.

Systems management uses quantitative and qualitative comparisons of alternatives to identify the one alternative best able to meet the objective. Assuming qualitative criteria provide a number of acceptable alternatives, quantitative techniques are then used to determine the most acceptable alternative.

Particular attention is focused on all the elements or subsystems associated with a given alternative. To ensure that each element is part of a viable alternative requires knowledge of its relationship to a given alternative.

A system involves a given set of relationships among input, process, output, and feedback design to provide an output that will meet the objective. The process is illustrated in Figure 5-7.

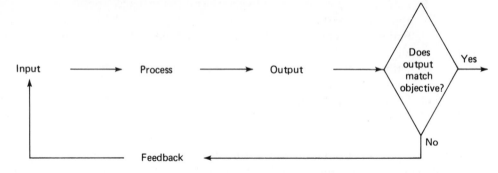

FIG. 5-7. Components of a System

In systems management several steps are performed:

1. Isolate problems.
2. Define alternatives.
3. Evaluate alternatives using cost, time, effectiveness, and risk.
4. Select alternatives.
5. Design the selected alternatives.
6. Implement alternatives.

Figure 5-8 shows a flow diagram demonstrating how each of these steps is involved in a complex problem dealing with environmental pollution. Note how the six steps are shown on the left margin embracing various individual elements. The important thing to note is the interrelationships with the various elements and the number of parallel decisions that must be made. The systems approach allows the engineering manager to make his decision in the light of all the factors influencing it; the result will be a better solution to the problem.

The systems approach to management of complex projects is a tool for the project manager either through matrix management or project management. It allows him to successfully plan and control complex projects made up of many subsystems. It gives him a method by which all the elements can be set aside and examined, alternatives can be identified and selected, and action can be implemented.

SUMMARY   Objectivity is important in achieving proper project selection but it is difficult to achieve. Project selection may be made by ranking projects according to their functional characteristics. Systems of quantifying the aspects have been developed. However, the opinions are somewhat subjective and executives frequently have different opinions. These have proved useful nonetheless in pointing up project weaknesses.

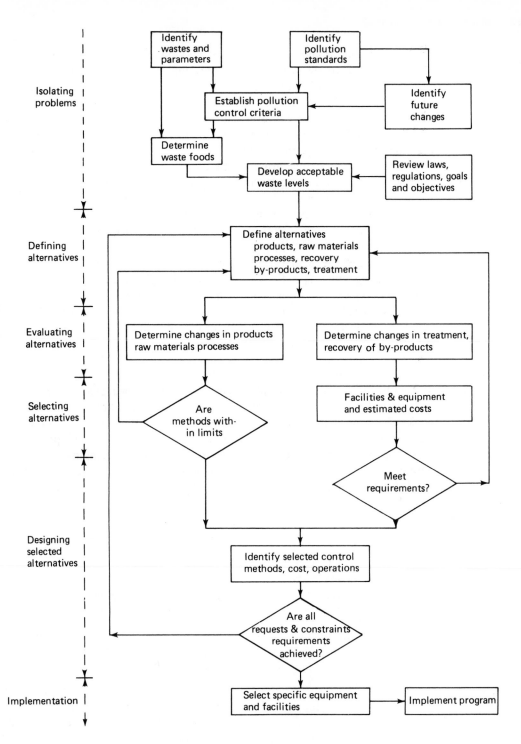

FIG. 5–8. System Approach to an Environmental Pollution Control Problem

An adjunct or alternative system of ranking involves the financial aspects of the project. Here the ranking is done through comparisons of cash flow, payment time, and rate of return on investment. This approach gets much closer to objectivity than the more subjective functional approach.

In the final analysis, however, each company has its own preferences, and it also has its own opinions on who should make such studies. The engineer manager may find he is not directly involved.

Matrix management provides the concentrated management effort that complex and unfamiliar undertakings demand. Matrix organization provides the opportunity for success in meeting schedules and staying within budgeted costs. However, such benefits are available to the organization only when all levels of management understand the unique features of matrix management and what is required to make it operate successfully.

Because of the great variety of projects and the relatively new concept of matrix management, specific concepts and principles are not yet formulated. Therefore, implementing a new position such as project manager into an existing organizational structure is a slow and sensitive process, especially when the position has significant power. Because of the complex interrelationships, the engineer manager must give special attention at the very beginning of the establishment of a matrix organization.

Matrix management provides better cost control, as more resources are applied to the projects. Jobs and budgets are better defined, changes better controlled, performance more closely watched, and action taken to prevent problems. The project manager's responsibilities may differ from one project to another, and they must be tailored to the particular needs of each project.

The matrix organization provides training for new employees and the development of functional specialization in the organization. Systems management provides support to project management in the most complex projects.

Important Terms

Functional Manager: Manager of a functional area of a corporation such as production, finance, or sales.

Matrix Organization: An organizational form that combines the functional and project forms into one organization structure.

Payout Period: The number of years required for cash earnings or savings generated by a proposed project to equal original capital investment.

Project Manager: Manager of a team working as a unit toward accomplishing a specific goal.

Ranking: Rating the various projects against one another by comparison of functional and financial aspects.

For Discussion

1. Why is objectivity important in project selection? Why is it difficult to obtain?
2. Describe the process of ranking by functional aspects.
3. Discuss the strengths and weaknesses of functional ranking.
4. Describe the process of ranking by financial aspects.
5. Discuss the strengths and weaknesses of financial ranking.
6. Why is rate of return an important criterion in project selection?
7. Why would each firm have its own way of ranking projects?
8. What are the characteristics of project management? Discuss its advantages and disadvantages.
9. What is the meaning of matrix organization or matrix management? Why is it used?
10. Discuss the attractiveness of matrix management.
11. Describe how you would design a matrix organization.
12. What situation dictates the need for project management?
13. Describe how you would establish a matrix organization.
14. Name seven things over which the project manager in a matrix organization must have control if he is to be effective.
15. Name six causes of failure of a matrix organization.
16. Define systems management. How can it be helpful in project management?

Case 5-1

Introducing the Matrix

Manchester Industries has grown rapidly in recent years. It is organized along functional lines which have worked well for its operations. However, each of its expansions has involved major facility installations and the firm has found it difficult to control these expansions with its present organizations.

Paul Johnson, the president, recently sent Alberta Jones, chief engineer, to an AMA conference on matrix management, a system for handling projects. Upon her return, Alberta suggested to Paul that matrix management might help the firm during this period of rapid expansion. Paul was impressed and suggested that on the two upcoming projects Alberta arrange to try this method of operation.

1. How should Alberta go about implementing matrix management?
2. What problems should Alberta anticipate?

Case 5–2

Meeting Deadlines

Rex Smith had assigned three of his best engineers to design and specify a new piece of equipment to handle the take off, cutting, and stacking of product from a new line. Conveyors, pull rolls, saws, and stackers would be involved. He assigned various segments of the job to each of the engineers. Jim was to secure information on conveyors, pull rolls, and saws and Mike was to get information on handling and stacking equipment; Joe was to develop the design drawings for the entire take off and handling system. They were given three weeks to have a report ready.

Rex called in the group for a progress report two weeks later. He found Jim had gotten information on conveyors, pull rolls, and saws and was off on another project. Mike had several quotes on stacking equipment. Joe was in the preliminary design stages for the entire system.

Rex listened patiently to their stories of progress, but with only one week to go, he was apprehensive about meeting the deadlines. He set up a somewhat more detailed schedule for each of them aimed toward meeting the completion by the following Friday. A schedule was typed and distributed.

The following Friday found the specifications still not completed, and little progress had been made.

1. What is wrong with Rex's approach to accomplishing a task?
2. Is there anything wrong with his follow-up?
3. How should he have approached the problem initially?
4. What should he do now?

For Further Reading

AMES, CHARLES B., "Payoff from Product Management," *Harvard Business Review* (November–December 1963), 141–52.

FULMER, ROBERT M., "Product Management: Panacea or Pandora's Box?" *California Management Review* (Summer 1965), 63–74.

GALBRAITH, JAY R., "Matrix Organization Designs," *Business Horizons* (February 1971), 29–40.

JACOBS, RICHARD A., "Project Management—A New Style for Success," *Advanced Management* (Autumn 1976), 4–14.

KARGER, DELMAR W., and ROBERT G. MURDICK, *Managing Engineering and Research.* New York: Industrial Press, 1963.

KOONTZ, HAROLD, et al., *Management.* New York: McGraw-Hill Book Company, 1980.

KOENIG, MICHAEL H., "Management Guide to Resource Scheduling," *Journal of Systems Management* (January 1978), 24–29.

LUCK, DAVID J., and THEODORE NOWAK, "Product Management—Vision Unfulfilled," *Harvard Business Review* (May–June 1965), 143–57.

McGINNIS, L. F., and H. L. W. NUTTLE, "Project Coordinator's Problem," *Omega,* 6, No. 4 (1978), 325–30.

MIDDLETON, C. J., "How to Set Up a Project Organization," *Harvard Business Review* (March–April 1967), 19–28.

PARK, WILLIAM R., *Cost Engineering Analysis.* New York: John Wiley & Sons, Inc., 1973.

ROLEFSON, JEROME F., "Project Management—Six Critical Steps," *Journal of Systems Management* (April 1978), 10–17.

TOELLNER, JOHN, "Project Estimating," *Journal of Systems Management* (May 1977), 6–9.

YOUKER, ROBERT, "Organization Alternatives for Project Managers," *Management Review* (November 1977), 46–53.

CREATING AN
EFFECTIVE TEAM

The engineer manager must become skilled in developing working relationships with people and must recognize that he is unable to be personally involved in all activities. He must learn to assign responsibilities and delegate requisite authority.

LEARNING
OBJECTIVES

☐ Be able to compare and contrast the various leadership theories discussed.

☐ Learn to develop effective leadership characteristics.

☐ Learn to assign responsibilities and delegate authority.

☐ Become familiar with manager's reluctance to delegate.

☐ Gain knowledge on how to make delegation effective.

EXECUTIVE COMMENT

MILTON R. GAEBLER
Senior Vice President — Engineering
Anheuser-Busch Companies, Inc.

Milton R. Gaebler is senior vice president of engineering for Anheuser-Busch, Inc. He received a B.S. in Mechanical Engineering from the University of Missouri-Columbia in 1940. After eight years in various engineering and supervisory positions with Monsanto he joined Anheuser-Busch in 1949 as senior plant engineer. In 1963 he became superintendent of projects and in 1970 he was promoted to chief engineer. Five years later he was appointed to vice president of engineering and then senior vice president.

Creating an Effective Team in an Engineering Organization

The engineering graduate embarking on his technical career has an intense desire to apply his newly acquired technical knowledge in the design of plants, equipment, and the solution of engineering problems. He personally executes his assignments to achieve the desired objective.

As the technical engineer progresses through the ranks of his engineering organization to the management level, he must not only retain his technical expertise but must also develop new skills, the skills of management that allow him to develop effective teams and thereby expand himself and his accomplishments.

The manager's success depends on his ability to select competent technical personnel as well as capable members of his management staff. Investment in continued training and enrichment of these personnel make possible the continuing success and growth of the organization.

The engineering manager who has a competent staff will find delegation of authority and responsibility a minimum risk. Many engineers because of their early experience in the profession are reluctant to delegate these authorities and responsibilities. To be successful, an engineering manager must develop and practice these characteristics.

The engineering manager is a leader who is continually motivating other people to accomplish tasks. The large organization that combines the services of many people requires direction. Direction involves relationships with people within the engineering department, within the company, and outside the company. It involves relationships at all levels: with people at the same level in the department and with people at higher and lower levels within the company.

Leaders may not be managers, but managers must be leaders. A good leader will induce his subordinates to work with confidence and zeal. Planning, organizing, and staffing alone will not get a job done. Direction is essential.

The engineering manager must communicate with clarity and integrity. Working with others requires the ability to transmit information verbally and in written form so that those who will be executing the work have a clear understanding of the objectives and goals.

The successful engineering manager always remembers that he is working through and with people to attain the objectives of the organization. The selection of staff, the structure of the organization (and within the organization structure the delegation of authority and responsibility), and direction and leadership are the prime ingredients of a successful organization. To these are added planning and control to complete the management function.

The largest project is no larger than its smallest part. The key to handling large projects is to divide them into many small, logical, and coherent projects, each manageable by an individual. Thus the project manager works through and with people to design and build large facilities. He does not do it alone.

The engineering manager who early realizes that he personally cannot be involved in all details and learns how to work through others will multiply himself a thousandfold.

The engineer has an aptitude for working with machines and physical materials. He is an innovator in every respect and is able to put things together and make them work. After the engineer develops or understands a relationship, he expects it to work the same way the next time, because he is familiar with laws and theories concerning physical relationships. As an engineer, he is always personally involved in mechanical activities; in fact, he is really most comfortable when he is working directly with machines and equipment.

As an engineer manager, he has similar involvement and relationships, but these relationships are not with physical things that always react the same; they are with people who may react differently each time. There will also be a wide range of people reactions with which he was unfamiliar as an engineer. There are, however, engineering management concepts and principles that he can utilize to create effective relationships in order to accomplish his goals or objectives through others.

ACCOMPLISHING GOALS OR OBJECTIVES THROUGH OTHERS

When he worked as an engineer, he accomplished his goals or objectives by himself or at most with one or two other engineers. One reason he enjoyed his work so much was that he worked by himself, could set a goal, and then work to accomplish it. He was always efficient in everything he tried to do and consequently was successful as an engineer. Because of this success, he is now an engineer manager with engineering management responsibilities. He must now realize that as one person he is limited in what he can do. He cannot personally do all the tasks necessary to accomplish the prescribed goals and objectives. As his organization grows, it is impossible for him to make all decisions, perform all tasks, and be directly involved in all activities. Because of these restraints, and because he lacks the time, he must work through people to get things done. His role in the organization is to get things accomplished through others, not by himself. This allows the engineer manager to accomplish projects that by himself he could not achieve.

The engineer manager has established goals or objectives for the activities for which he is responsible and has prepared plans, policies, procedures, and rules to accomplish these. He transforms goals or objectives into meaningful terms which are comprehensible to the individual. His planning efforts deal with some of the major considerations in deciding between alternative courses of action. He decides in advance the what, how, when, and who. These decisions involve coordinating the plans that are developed, communicating them to his people, and obtaining their participation to make the plans successful.

To ensure that these plans are successfully implemented and to effectively accomplish the goals, the engineer manager designs an organizational structure. In organizing this structure, he develops roles for people to fill, plans the major activities or duties to be performed, and organizes the relationships that will be involved. The structure will allow the group to cooperate and work together most effectively, for each member will know the role to perform. The need for developing an organizational structure grows out of the need for cooperation to achieve the planned organization goals as well as the personal goals of the participating individuals. Through this organizing effort the engineer manager brings order and creates an environment of cooperation among the members of his group.

Leadership is an important aspect of managing and therefore important in creating an effective team. By exercising the proper leadership, the manger creates followers who perceive that he can provide the means for achieving their own desires, wants, and needs. So we might define leadership as the art or process of influencing people so that they will strive willingly toward the achievement of group goals.

Katz and Kahn defined leadership as "the influential increment over and above mechanical compliance with the routine directives of the organization."[1] In other words, it is that something extra that causes the individual to rise above the specific requirements of his position to follow a manager towards the overall goal with zeal and enthusiasm. This clearly underlines what separates the effectiveness of one department from another.

It can perhaps be said that few subordinates work with continuing zeal and competence. It is probably not the nature of most human beings to do so. It is a role of the leader, however, to cause the average individual to see the organizational goal and willingly follow him in achieving it.

Nature of Leadership

Every group of people that performs near its total capability has some person in charge who is skilled in the art of leadership. Koontz, O'Donnell, and Weihrich state that "this skill seems to be a compound of at least three major ingredients—the ability to comprehend that human values have differing motivating forces at varying times and in different situations, the ability to inspire, and the ability to act in a way that will develop a climate for responding to and arousing motivations."[2]

The first of these involves motivation theory and will be dealt with in some depth in the next chapter. While managing is situational and dependent on contingencies, if a manager is aware of motivation theory, he is better able to satisfy the needs of his people and hence secure the desired responses.

The second ingredient, the ability to inspire, is more difficult to define. Some seek charismatic qualities that induce loyalty and devotion.

The third ingredient has something to do with style of leadership and is mentioned later in this chapter. Organizational climate tends to influence the willingness to follow. It is therefore important to provide the proper environment in order to get the proper performance.

The Trait Approach to Leadership

It was originally thought that a leader could be identified by traits and that if we could find these traits in an individual, we could be assured that he would become a leader. Those who believed in this approach felt that leaders were born, not made.

[1]D. Katz and R. L. Kahn, *The Social Psychology of Organization,* 2nd ed. (New York: John Wiley, 1978), Chap. 16.

[2]H. Koontz, C. O'Donnell, and H. Weihrich, *Management,* 7th ed. (New York: McGraw-Hill, 1980), p. 663.

Prior to 1949 most studies of leaders did tend to concentrate on identifying leadership traits. Stogdill found that various researchers identified five physical traits related to leadership ability (such as energy, appearance, and height), four intelligence and ability traits, sixteen personality traits (such as adaptability, aggressiveness, enthusiasm, and self-confidence), six task-related characteristics (such as achievement, drive, persistence, and initiative), and nine social characteristics (such as cooperativeness, interpersonal skills, and administrative ability).[3]

However, the study of these traits has not been very helpful in identifying leaders. Some leaders possess the traits and many nonleaders possess them. Furthermore, conflicting data have been developed in which there is a considerable lack of uniformity. Stogdill did find that there are a few characteristics that may have some degree of correlation, including intelligence, scholarship, dependability, responsibility, and social participation.[4] Ghiselli found significant correlation between leadership ability and such traits as intelligence, initiative, and self-assurance.[5]

The Situational Approach to Leadership

When it was found that the trait approach was not the answer to selecting leaders, attention turned to the situational approach. It was felt that perhaps leaders were a product of a given situation. Hitler of Germany, Mussolini of Italy, and Franklin Delano Roosevelt of the United States were considered examples. We referred earlier to the fact that followers tend to look for leaders who will help them achieve their own personal goals. In all three of these cases, certainly the followers did perceive this result.

Other studies have shown that effective leadership does seem to depend on environmental factors as evidenced by situations. It does not help much in the selection of leaders, perhaps, but it does mean that a manager's performance may be affected by the way the environment is designed.

Fiedler's Contingency Approach to Leadership

Fiedler and his associates at the University of Illinois have combined the trait and situational approaches to leadership and suggested a contingency theory. The Fielder model proposes that effective group performance depends on the proper match between the leader's style of interacting with subordinates and the degree to which the situation gives control and influence. He developed an instrument called the *Least-Preferred Co-Worker* (LPC) questionnaire that purports to measure whether a person is task- or relationship-oriented. He isolated three situational criteria—leader–member relations, task structure, and position power—that he believes can be manipulated so as to create a proper match with the behavioral orientation of the leader.

[3]R. M. Stogdill, *Book of Leadership* (New York: The Free Press, 1974), pp. 74–75.

[4]R. M. Stogdill, "Personal Factors Associated with Leadership: A Survey of the Literature," *Journal of Psychology,* 25 (1948), 35–71.

[5]E. E. Ghiselli, "Managerial Talent," *American Psychologist* (October 1963), 631–41.

Fiedler found three "critical dimensions" of the situation that affected the leader's performance:

1. *Position power*. This represents a degree to which the power of position, as distinguished from charismatic or expertise power, enables the leader to secure compliance with directions. Stated another way, this may be the power arising from organizational authority. Fiedler points out that the leader who has clear position power can more easily obtain followership than one who does not have such power.

2. *Task structure*. With this dimension, Fiedler referred to the extent to which tasks can be clearly spelled out and the people as a result held responsible, in contrast to those situations in which tasks are vague and unstructured. If tasks are clear, the quality of performance can be more easily controlled.

3. *Leader–member relations*. Fiedler regards this dimension as the most important from a leader's point of view, since position power and task structure may be largely controlled by the organizational design. This item has to do with the extent to which members like and trust the leader and are willing to follow him.

Some important conclusions of Fiedler are:

Leadership performance depends then as much on the organization as it does on the leader's own attributes. Except perhaps with the unusual case, it is simply not meaningful to speak of an effective leader or an ineffective leader; we can only speak of a leader who tends to be effective in one situation and ineffective in another. If we wish to increase our organization and group effectiveness, we must learn not only how to train leaders more effectively, but also how to build an organizational environment in which the leader can perform well.[6]

The Managerial Grid®

One of the most widely known approaches to dramatizing leadership styles is the Managerial Grid developed some years ago by Robert Blake and Jane Mouton.[7] This Grid, which is shown in Figure 6–1, has been used widely throughout the world as a means of managerial training and a way of identifying various combinations of leadership style.

Blake and Mouton describe five basic styles. Under the 1,1 style, managers concern themselves very little with either people or production and have minimum involvement in their job. They act primarily as messengers communicating information from superiors to subordinates. The 9,9 managers display in their actions the highest possible dedication both to people and to production. They are the real "team managers" who are able to integrate the production needs of the enterprise with the individuals.

The third style is identified as 1,9 management, in which managers have little or no concern for production but are concerned only with people. They promote an environment that is relaxed, friendly, and happy, but one in which the people are not concerned with putting forth a coordinated effort to accomplish a

[6]F. E. Fiedler, *A Theory of Leadership Effectiveness* (New York: McGraw-Hill, 1967), p. 261.

[7]Robert R. Blake and Jane S. Mouton, *The Managerial Grid* (Houston: Gulf Publishing Company, 1964), p. 10.

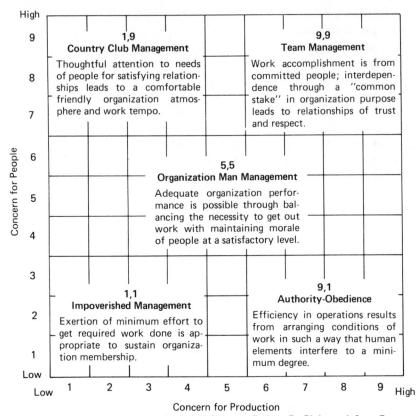

FIG. 6-1. The Managerial Grid® [Source: Robert R. Blake and Jane S. Mouton, *The New Managerial Grid*, (Houston, Gulf Publishing Co., 1978.)]

defined goal. The fourth style is the 9,1 manager, who is concerned only with developing an efficient operation and has little or no concern for people.

In the middle of the Grid is the 5,5 orientation. Clearly, 5,5 managers would have a balanced concern for both production and people. This, however, can represent primarily mediocre performance. This is a convenient way of classifying managerial styles, but it does not tell us why a manager falls in one part or another of the Grid. To find this out, a person has to identify underlying assumptions.

DEVELOPING LEADERSHIP CHARACTERISTICS   It can be seen from the preceding review of leadership theories, that one of the most important characteristics for the engineer manager to develop is to be an effective leader. Being a successful leader involves a continual effort to influence the behavior of others. He achieves leadership by making sound proposals and decisions over a period of time. As a result, he gains the confidence of his subordinates.

Leadership involves helping subordinates use their capabilities to achieve the goals of the group. Such involvement requires the leader to give full cooperation and encouragement—not create rules and regulations that prevent members of the group from achieving them. Through this effort, morale will be high and the group will achieve greater efficiency, resulting in a strong and successful organization.

The engineer manager's position gives him vested formal authority, but to be effective, he strives not to use this authority. Rather, he seeks to lead his subordinates by developing clear goals or objectives and by leading the group to achieve them. This helps to satisfy their personal needs, and it creates an atmosphere of respect for him as leader.

People need someone to follow. Most people find it is much easier to follow those who can satisfy their personal goals. They are in turn motivated when their needs are met. A fundamental concept is that people follow those who can fulfill and satisfy their needs most effectively.

A leader must be enthusiastic, for his own behavior must serve as an example for the group to follow. For example, when the engineer manager is open and honest with his subordinates, they will tend to be open and honest with him; when he demonstrates initiative, they will show initiative; when he is friendly and cooperative, they will respond accordingly.

He must periodically reevaluate his attitudes and actions to discover those which may have a negative influence on his subordinates and he must take corrective action. He must develop a habit of watching the favorable and unfavorable reactions to his subordinates. He must always realize that people do not react consistently. Consequently, to be an effective leader, the engineer manager must know his people.

Corrective Action

An engineer manager must develop the ability to recommend corrective action to his subordinates so that they will not be offended by his remarks, but will instead feel good and want to work all the harder. This requires him to understand his subordinates' feelings and attitudes and to know something of their personal lives. He must strive to understand thoroughly the cause of the problem. When recommending corrective action, he must use a positive approach, first recognizing and complimenting the subordinates on good performance and then moving to the areas that need corrective action. This reminds the subordinates that good performance is rewarded by recognition and it encourages them to want to correct those areas of performance in which they have weaknesses.

Integrity

The engineer manager must develop a reputation for being trustworthy. He must be completely honest with himself. Subordinates quickly recognize one who is deceitful or who makes promises but has no intention of following through. Confidence in him by his superiors and subordinates is essential. They must be able to take him at his word—to count on him to perform. If he is consistent, subordinates will develop a feeling of confidence in him. When the engineer manager is trustworthy and reliable, his subordinates will recognize him as their leader and will be proud and happy to follow him. When subordinates lose this confidence, he soon loses his ability to lead.

Keeping Informed

To be an effective leader, the engineer manager must keep informed of events, both inside and outside the organization. This requires him to recognize and understand what his subordinates are telling him. Their advice helps him to know what they are thinking. It helps him to select goals that they will welcome and that will be meaningful to them. By this means, he gains loyalty and respect.

Obtaining information on what is happening both inside and outside the organization takes time and effort on his part. He must have much contact with his subordinates before they will confide their ideas to him. Interaction with his subordinates will help him to become an effective leader.

Group Participation

Subordinates like to initiate and participate in action to accomplish their tasks. Participation keeps the subordinates involved and informed rather than allowing them to learn about events from others. It is best used when subordinates are involved in nonroutine activities in which they can make a contribution and become more confident in their ability and skills.

STYLES OF LEADERSHIP

Engineer managers can be classified according to their style of leadership: autocratic, diplomatic, consultative, or participative. The autocrat maintains authority through his position; he demands power. The diplomat is a salesman; his authority is achieved through personality and he generally works toward obtaining power. The consultant receives authority by his reputation and simply accepts power. The participant gains authority through his competence and usually rejects power. These four styles of leadership can be summarized as one who tells, one who sells, one who consults, and one who joins. The engineer manager must realize that, regardless of his style, he must not force it on his subordinates. They may be much more effective operating in the environment of another style. The engineer manager wants his people to complement him, not duplicate him. Items influencing his particular style of leadership include his personality, his attitude toward work and people, and his own needs and goals.

The engineer manager must learn to adapt and change his style to fit the environment in which he operates. This means that he must use several styles, depending on his position and immediate problems (Figure 6–2). He must be careful that he does not change his style too quickly or vary it from day to day. Such inconsistency will generally cause his subordinates to lose respect for him as a leader. He may change his style when he changes positions and matures with age and experience. The important point is that an effective leader has a range of leadership styles in which he is most comfortable and effective.

Factors Determining Effective Leadership Styles

From our study of leadership research we might conclude that the factors that determine the most effective style for the engineer manager are:

1. The expectations of his subordinates, their degree of dependence on him, and their experience
2. The engineer manager's own personality and attitude toward the various styles
3. The styles his superiors practice and the management philosphy of the company
4. The level of the engineer manager's position in the organizational structure and his responsibilities

In summary, the engineer manager must realize that he will need to change his style of leadership to fit his position and environment.

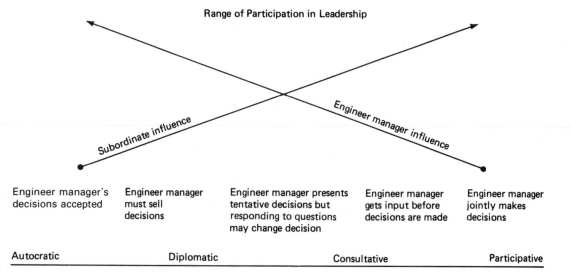

FIG. 6–2. Leadership Styles the Engineer Manager May Use

AUTHORITY AND RESPONSIBILITY   Authority is related to one's position within the organization. Each administrative position has certain rights that the incumbent acquires upon assuming that position's rank or title. It has nothing to do directly with the person. Authority lies in the position. The opposite can likely be said, that as soon as one vacates the position the inherent rights are lost.

In developing an organizational structure, the engineer manager forms superior–subordinate relationships. These relationships establish the concept of authority which gives meaning to the engineer manager's position in relation to others and gives him formal authority. It should be noted that he also acquires informal authority, which many consider the only real authority. This is obtained by his gaining respect and recognition from his subordinates.

These authority relationships tie the various organizational units together. Such authority is both vertical (the superior–subordinate relationships) and horizontal (the relationships between management of the same level and with staff units in the department). To the engineer manager, these relationships involve not only people but also physical resources involved in accomplishing planned goals.

Authority gives the engineer manager the privilege to act or to require others to act. He acts to achieve the organizational goals or objectives through others, including making decisions and implementing them (Table 6–1). In achieving them through others he can use motivation, requests, threats, and force. His position carries the right to direct people and resources, request subordinates to perform specific assignments, and motivate them to accomplish the tasks. The real authority comes from the subordinates' acceptance of the engineer manager as their leader. This acceptance depends on his ability to obtain support from his subordinates through their desire to achieve the organizational goals. He has many techniques to obtain this support, including power to grant and/or withhold rewards or to dismiss. A very important source of authority is the engineer manager's personal ability in engineering management. This recog-

---

TABLE 6–1

**Factors that Determine the Number of Employees
An Engineer Manager Can Effectively Supervise**

Frequency of superior/subordinate contacts
Training of superior/subordinates
Type of work
Maturity and motivational level of subordinates
Engineer manager's ability to plan and organize
Clarity of policies
Engineer manager's communication skills

nition arises from his understanding of problem areas and ability to provide sound technical and management answers. Authority achieved in this manner will be accepted not only by his own subordinates but also by others outside the department.

Most engineer managers are very concerned about their "limits" of authority, as defined by company policies, work rules, and procedures. They become involved in ensuring that subordinates follow these rules completely and in detail. When this happens, these engineer managers do not feel they have any authority, even to make minor decisions. In most cases, there are many more things they can do within the limitations of policies, rules, and procedures. It is important for the engineer manager to list for himself those things he does have the authority to do. He must get to the point where he is managing, not merely monitoring actions.

Delegation makes it possible to group activities into sub-divisions to effectively achieve goals. It is seldom that the engineer manager does not understand the role of delegation involved in the delegation process. Delegation, however, is the least well-practiced of any engineering management concept.

THE DELEGATION PROCESS

Frequently, lip service is given to it, but there is failure to actually put it into action. The engineer manager does this at his peril, however, for delegation plays a major role in the management of his time.

When activities are grouped into departments or subdivisions to facilitate the accomplishment of goals, the manager of each subdivision must have the authority to coordinate its activities with the organization as a whole. The more a subdivision or department has a clear definition of the results it is expected to achieve, the activities to be undertaken, the authority delegated to it, and the relationship with other units, the more adequately the individuals in the various subdivisions can contribute toward accomplishing the overall objectives. To define a job and delegate authority to do the job requires patience, intelligence, and clarity of objectives. It also requires that the manager doing this understands clearly the responsibilities that he must exercise.

The delegation process involves three important phases that cannot be separated. They are:

1. *Assignment* of duties in terms of the results the engineer manager wants achieved. Each person in the organization has some part of the total job to perform.

2. *Authority* to carry out the assignment. This is the right to decide about resource allocation and utilization. The limits of authority must be clearly defined within company policy.

3. *Accountability* to accomplish the assignments or to perform the duties assigned. When one accepts an assignment, he is in effect making a promise to do his best to accomplish it. It is his responsibility to perform the job in the manner expected by his superior.

In the delegating process, authority must be granted when duties are assigned, and accountability is simultaneously exacted for the accomplishment of these duties. The engineer manager must realize that it is impossible to split this process; therefore, it is logically consistent that the authority needed to perform the assignment must be equal to the responsibility. In many cases, however, the engineer manager may fail by assigning a responsibility but not delegating the necessary authority to accomplish the task.

The flow of responsibility within an organization, therefore, usually begins with an assignment. This must be accompanied by the authority to ensure that the work is done. The authority may be formal or informal, as stated earlier. The purpose of assignment of responsibility is to relieve the engineer manager of many time-consuming duties and to allow him more time to complete his management tasks. Accountability is, however, one phase within the flow of responsibility that cannot be delegated downward. For example, the president of a company may assign many responsibilities throughout the organization, along with the authority necessary to accomplish the desired goals, but he alone is accountable to the board of directors. He is expected to discharge the responsibilities of his office efficiently and he is held accountable for the final results.[8]

It takes courage to delegate authority because such delegation does not absolve the engineer manager from responsibility. Being ultimately accountable may actually prevent delegation for fear of the failure of subordinates to whom he has granted authority and assigned responsibility.

Fortunately, successful delegation practices can be learned by the engineer manager and such practices can be taught to his subordinates. The ideal time to start this training is during the first supervisory assignment. Even after the engineer manager has developed successful delegation techniques, he must constantly reexamine them or they will soon fail to be effective. He must rigidly discipline himself in his delegation practices. Without this kind of discipline, he will not only fail in his current position, but he will also carry the same bad habits with him to a new position.

RELUCTANCE TOWARD DELEGATION

An engineer manager must realize that, in practice, personal adjustments toward people are necessary in order to make delegation real and effective. These adjustments include overcoming the following feelings:

1. Reluctance to let his subordinates *make appropriate decisions* because they may not be exactly those the engineer manager would make. He must welcome the ideas of his subordinates and develop and cultivate their creativity. In many

[8]Melvin Silverman, *Project Management Organization* (New York: John Wiley, 1976), p. 5.

cases, the engineer manager's lack of confidence in a subordinate's ability is used as an excuse for not delegating. Typical explanations are "He needs more experience" or "He has excellent ideas but does not follow through." These may be valid reasons until the subordinate is properly trained, but they have no viability over the long term.

2. Reluctance to *trust* subordinates. Engineer managers have no other alternative but to trust their subordinates. Delegation implies a trust between the engineer manager and his subordinates. Because he is vesting them with his authority to represent him, this involves taking risks. This is part of the engineer manager's responsibility. This is what he is hired for and it is the only way he can be productive.

3. Reluctance to let subordinates *do their work themselves.* Many engineer managers believe they can "do it better themselves," which is true, because otherwise they would not have been promoted. But as engineer managers, they must work through others. This means taking time to train and develop subordinates, *not* doing it themselves. This is especially true in the specific technical area from which the engineer manager was promoted. There is a strong tendency to continue to be too interested and involved in the area he has left. It is difficult to discontinue making decisions for the person who has taken his old position. To be effective, the engineer manager must realize that he must change; otherwise, the very purpose of delegation is defeated. He must concentrate on the tasks for which he is now responsible.

4. Reluctance to let subordinates *make errors.* Everyone will make errors and misjudge situations, even the most successful engineer manager; therefore, he must realize that his subordinates will make errors. No one is saying that the engineer manager should ignore error after error. Continual checking of subordinates' decisions, however, and acting to ensure that no errors occur are impossible and thus negate delegation. The engineer manager can minimize such errors by allowing the subordinate to assume his delegated role and to make decisions in his area. When delegation is practiced, the engineer manager must counsel, train, and work with subordinates until he feels they can adequately assume their roles.

5. Reluctance to develop *broad controls.* When there is delegation, it is important to establish broad controls, so that at all times the engineer manager can determine whether or not the goals are being achieved. He should not delegate unless he is assured that the assigned responsibilities of his subordinates support the organization's goals. It must be noted that these are "broad" controls, not controls that involve constant checking of every detailed operation. Most engineer managers have difficulty accepting the importance of having "only" broad controls.

Certainly the engineer manager can be trained in delegation concepts, but he must understand and overcome these reluctances toward delegation. No one can tell him to trust people; he must do this himself. He must understand that this is a necessary part of delegation. Until he accepts these concepts in a very true sense—in fact, until they become a part of his lifestyle—he cannot be a successful engineer manager. Promotion to this position without fully understanding and accepting these concepts will produce an ineffective, unhappy, and dissatisfied individual.

CONCEPTS OF DELEGATION When the engineer manager makes an assignment, he should expect to spend considerable time training his subordinate. Training a subordinate to handle delegated authority involves a large amount of time in the beginning, but it will require a smaller amount of time over a long period. Since every engineer manager is always busy, this principle is difficult to understand and accept. Other concepts that will improve the engineer manager's delegation effectiveness are:

1. Authority delegated must be commensurate with the assigned responsibility. He must not hold people responsible for carrying out duties for which he has not delegated requisite authority.

2. There must be an effective and appropriate organization of activities. When relationships are clearly defined, the proper environment for effective delegation is provided. It avoids confusion among the various organizational levels that are receiving delegation. When the engineer manager has developed a meaningful organizational structure, many conflicts from overlapping or haphazard arrangements of duties are eliminated.

3. There must be direct superior–subordinate relationships, so that each person knows who is responsible for the delegated assignments.

4. It must be determined who should delegate for effective overall performance.

5. A subordinate must report to *only one superior*. Being responsible to more than one person causes conflicts for both subordinate and superior.

Making Delegation Effective

It is the engineer manager's responsibility to make delegation work in the organization. He must develop an environment for delegation and use accepted concepts to make it effective. It is up to him to make these concepts real to his subordinates. There are many ways that he may err, e.g., through partial delegation, inconsistent assignments, and making decisions for subordinates. These are all serious, but they can be corrected if the engineer manager will do the following:

1. Clearly define assignments in the light of results expected.

2. Grant the authority necessary to make possible the accomplishment of the assignments.

3. Consider ability and qualifications of subordinates when delegating, since these will influence the nature of the assignment. The good organizer will approach delegation primarily from the standpoint of the task to be accomplished.

4. Provide adequate instruction, guidance, and training for subordinates who are to be delegated authority.

5. Make himself available for counseling with subordinates after assignments are made. Because plans change and decisions must be made in the light of changing conditions, delegation tends to be fluid and to be given meaning in the light of such changes. This means there should be a free flow of information between superior and subordinate; it should furnish the subordinate information with which to make decisions and to interpret properly the authority delegated.

6. Design controls that show when problems occur. Since no manager can relinquish responsibility, delegation should be accompanied by techniques to make sure the authority is properly used; however, these controls must be relatively broad and designed to show deviation from the plan rather than to interfere with the detailed action of the subordinates. Managers should be ever watchful for some means of rewarding both effective delegation and effective assumption of authority.

7. Reward effective performance. The rewards need not be financial; they might have to do with granting greater discretionary powers or putting the recipient in a position for earlier promotion to a higher level.

Resistance to Delegation by Subordinates

In the delegation process the engineer manager not only delegates authority, but his subordinates also agree to accept it. There may be good reasons why the engineer manager's subordinates resist delegation; in fact, most of these reasons are because of past improper delegation practices of their superiors. Consequently, the subordinates have developed a pattern of doing everything to resist delegated authority. When the engineer manager allows this situation to occur, they will never accept this authority. To avoid such a relationship, the engineer manager must be certain that he is constantly aware of the reasons why subordinates resist delegation. Some of these reasons are:

1. *Taking the path of least resistance.* It is so much more convenient for the subordinate to ask the engineer manager for his solution to a particular problem, because the manager seems to enjoy making the decision for his subordinate and is always happy to respond to such requests. Therefore, all the subordinate needs to do is ask the engineer manager when an unusual situation develops.

Subordinates find that making good decisions themselves, especially at first, is hard work. It soon becomes a habit to depend on the engineer manager for his decisions.

Resistance to delegation develops because the engineer manager has not properly worked with and trained his subordinates to assume responsibility and probably has not mastered the delegation process himself. The engineer manager must understand that there is a great difference between training and coaching a subordinate and making decisions for him. Certainly, the engineer manager should always be available and happy to support and assist his subordinates, but it should be in an educational role, not doing their jobs for them.

2. *Fear of criticism.* Subordinates resist delegation if the engineer manager is unreasonable when they make errors or mistakes. People do not want to be ridiculed, especially in front of others. In fact, people resent and will even react sharply to such actions. Many times the subordinate gets the blame for something that went wrong which was beyond his control. If he is exposed to such criticism, he will then use every effort to sidestep accepting subsequent delegation.

When the engineer manager reviews the situation and gives constructive comments, the subordinate usually accepts them. In many cases. the subordinate readily accepts these comments because he already realizes he was in error.

3. *Lack of training.* When subordinates do not have what they believe to be adequate training, they feel uncomfortable in accepting an assignment. In this kind of an environment, the engineer manager's subordinates tend to reject delegated authority.

4. *Lack of adequate recognition.* The engineer manager must make every effort to ensure that his subordinates receive appropriate rewards, such as pay increases, better opportunities, status recognition, or other rewards for outstanding performance. However, he must realize that each subordinate views rewards differently and must be provided with positive incentives that are important to him. A reward that is effective for one group or individual may be ineffective for another group or individual. Also, attitudes toward rewards change from time to time. For example, recognition of status may be very important at one time to an individual, but at another time financial rewards may be most important. The engineer manager must realize that it takes effort to understand and determine what inducements an individual subordinate will respond to and consider important.

Thus, a variety of reasons cause subordinates to resist having authority delegated to them. Fortunately for the engineer manager, such resistance by subordinates can be overcome through effective delegation techniques. When the engineer manager first enters the position, his subordinates may resist the authority which he delegates to them. They are unsure of how he will react when problems occur. The engineer manager must realize this and spend sufficient time in training his subordinates to receive delegated authority.

To help the engineer manager recognize his delegation problems, he should ask himself the following questions:

1. Does he work on weekends and evenings more than others do?
2. Is he constantly interrupted by questions from his subordinates?
3. Is he always late in meeting deadlines?
4. Does he lack time for planning?
5. Is he always rushed, never having time to make decisions?
6. Does he demand details from subordinates?
7. Does he enjoy details and routine work?
8. Does he make decisions others should be making?
9. Are appointments impossible to obtain?
10. Does he get information for decisions from his immediate staff and secretaries rather than from the line?
11. Does he want to be involved in everything, especially details?
12. Does he withhold information from subordinates?
13. Does he obtain ideas from subordinates?
14. Is he afraid of subordinates gaining recognition?
15. Is there no one available to take over when he is absent from the job?
16. Does he lack time for community activities?
17. Does he maintain a tight organization?
18. Does the group seem to lack initiative or drive?
19. Do subordinates appear to be uninterested and dissatisfied with their work and the company?
20. Are subordinates not allowed to make decisions?
21. Do subordinates lack creativity?
22. Is there a team approach to problem solving?

When the engineer manager finds himself answering "yes" to a majority of these questions, he has probably failed in some way in his delegation practices. It is important for him to review carefully the principles involved in delegation and his training program with subordinates.

SUMMARY

When an engineer becomes a manager, he can no longer personally do all the tasks necessary to accomplish the prescribed goals or objectives. He must therefore organize his subordinates and work through them in accomplishing these goals. He should develop an organizational structure that will identify the roles that people will fill and plan the major activities necessary to meet the goals.

In developing an effective team, he must be a leader. Leadership has been defined as the art or process of influencing people so that they will strive willingly toward the achievement of group goals. Early efforts to find out what makes a good leader centered around a study of traits, but modern research tends to point toward a situational approach. Leadership performance seems to depend as much on the organization and its environment as on the leaders' own attributes. The Blake and Mouton managerial grid provides a convenient way of cataloging styles of management.

From time to time the engineer manager must reevaluate his attitudes and actions, looking for any negative influences on his leadership ability. He should keep his group informed of his expectations and assist them in achieving their goals.

To effectively reach the designated goals from his position in management, he must assign responsibility and delegate the requisite authority. These assignments should be made very clear and a satisfactory control system developed for follow-up. The manager should review any reluctance he has to delegation because delegation is so important to his success. He also needs to train his people to accept the delegated authority and to carry out the duties they have been assigned. He needs to be perceptive of their feelings that may make them reluctant to accept delegated authority.

Important Terms

Authority:	Authority is related to one's position within the organization and represents the degree of discretion conferred upon him to make it possible for him to fulfill his assigned responsibilities.
Delegation:	The act of conferring upon a subordinate the authority necessary for him to accomplish his assigned responsibilities.
Fiedler's Model of Leadership:	Fiedler's model indicates that leadership performance depends as much on the organization and its environment as it depends on the leader's own attributes.
Leadership:	Leadership is the art or process of influencing people so that they will strive willingly toward the achievement of group goals.
Managerial Grid:	An effective way of dramatizing leadership styles developed by Robert Blake and Jane Mouton.

For Discussion

1. Select two of the styles of leadership (autocratic, diplomatic, consultative, or participative) and discuss how leaders with these styles would handle a typical business situation.
2. Since it is so important for an engineer manager to have the ability to delegate and to trust his subordinates, describe how a company could evaluate this ability before advancement.
3. What are some of the problems involved in delegating authority, both from the engineer manager's viewpoint and the subordinate's viewpoint? What can the engineer manager do to avoid these problems?
4. Discuss the following statement in detail: "In the delegation process, the engineer manager not only delegates but his subordinates also agree to accept."
5. Why must the engineer manager work through people to get things done?
6. Why does the engineer manager design an organizational structure?
7. How can an engineer manager become an effective leader?
8. Comment on the following statement: "All an engineer manager really needs is formal authority."
9. The delegation process involves three important phases. List and discuss each phase.
10. What can the engineer manager do to make delegation real to the subordinate?
11. List and discuss some of the reasons why subordinates resist delegation.
12. What are the most important characteristics for a successful engineer manager? What should he do to develop these important characteristics?
13. What factors cause an engineer manager to resist delegating properly?
14. Why is it important for a manager on the job to keep up with the personal needs of his people?
15. Should an engineer manager always use the same "style"? Explain.

Case 6–1

The Overworked Engineer Manager

Rex Smith was the only engineer in the organization eight years ago. As the firm grew, additional engineers were employed and Rex became their supervisor. Because the department was small Rex continued to work along with the engineers even though he was the supervisor. Two years ago Rex found he could no longer handle the paperwork and scheduling and he hired Sally Carr as secretary.

More and more he found himself asking Sally to make decisions about the office, e.g. ordering supplies, setting up appointments, and scheduling projects. Rex continued to spend most of his time working on projects. If a difficult project came up, Rex would do the work himself.

Because of the demands on Rex's time, Sally began to take over more responsibilities, including setting up job assignments and handling questions from the engineers. In fact, when she made weekly work schedules, there was less confusion than when Rex tried to do it while continuing to work on projects.

One day Sally approached Rex with a new way to schedule projects and keep track of them as they progressed. Rex listened and commented, "That's a good idea, but you should spend your time being the secretary. I am the manager here and keeping track of projects is my responsibility." Sally replied, "I don't care what you call yourself; I'm doing more to make this department run smoothly than you are. You act more like an engineer than an engineer manager."

"You must be joking," Rex replied, "I am the manager and you are the secretary for the department."

"I've been taking on more and more of the responsibility here and I've been hoping that you would give me the title of 'Administrative Assistant,'" said Sally.

"The engineers accept your schedules because of me. If I were not here, nothing would happen. They get the job done because they know I am the boss and their leader," proclaimed Rex.

"Let's not argue about it, Rex. If you want me to be just a secretary, I'll stop managing," said Sally. "But if I do, you will have to stop most of the project work and spend your time managing the department."

1. What caused this situation to develop?
2. What managerial functions was Sally performing?
3. What fundamental error was Rex making?
4. Do you think Sally should be appointed administrative assistant?
5. What should Rex do now?

Case 6-2

The Art of Delegating

As Rex Smith's work grew in R&D he felt the necessity of appointing another engineering supervisor. He chose Pete Hammond, one of his best engineers, who took the job somewhat reluctantly because he was not sure he wanted to be in a managerial position.

Once having accepted the job, however, Pete felt he must make a success of it since he had been successful as an engineer. He remembered having read in a management book that one of the great failings of the new engineer manager is that he continues to do the engineering work himself. He doesn't delegate. Pete vowed to himself that this wouldn't happen to him.

Pete soon found it wasn't as easy to do as the book said. He had trouble getting the engineers in his group to take responsibility and he found himself checking and redoing some of their work. But if he didn't delegate, he found there was a bottleneck and the work didn't get done. Deadlines were being missed, and morale was sagging.

Pete thought he had a good group of people and wondered what he was doing wrong. He thought perhaps he should talk to Rex.

1. What is Pete doing wrong?
2. At this stage what can he do to improve matters?

For Further Reading

ALLEN, LOUIS A., "How to Stop Upward Delegation," *Nation's Business* (July 1973), 16–22.

BARNARD, CHESTER, "The Nature of Leadership," in *Organization and Management,* ed. Chester Barnard. Cambridge, Mass.: Harvard University Press, 1958.

BLAKE, ROBERT, and JANE S. MOUTON, *The New Managerial Grid.* Houston: Gulf Publishing Company, 1978.

BRECK, EDWARD, "Don't Hand Over Just Work, But Responsibility," *Business Administration* (May 1977), 28–30.

BURNS, J. M., *Leadership.* New York: Harper & Row Publishers, Inc., 1978.

FIEDLER, FRED, *A Theory of Leadership Effectiveness.* New York: McGraw-Hill Book Company, 1967.

FILLEY, A. C., R. J. HOUSE, and S. KERR, *Managerial Process and Organizational Behavior.* Glenview, Ill.: Scott Foresman & Co., 1976.

FISCH, GERALD G., "Do You Really Know How to Delegate?" *The Business Quarterly* (Autumn 1973), 17–20.

———, "Toward Effective Delegation," *CPA Journal* (July 1976), 66–67.

GHISELLI, E. E., *Explorations in Managerial Talent.* Santa Monica, Calif.: Goodyear Publishing Co., Inc., 1971.

GUEST, R., et al., *Organization Change Through Effective Leadership.* Englewood Cliffs, N.J.: Prentice-Hall, Inc., 1977.

HAYNES, MARION E., "Delegation: Key to Involvement," *Personnel Journal* (June 1974), 454–56.

HERSEY, PAUL and KENNETH BLANCHARD, *Management of Organizational Behavior.* Englewood Cliffs, N.J.: Prentice-Hall, Inc., 1977.

HOUSE, ROBERT, "A Path Goal Theory of Leader Effectiveness," *Administrative Science Quarterly,* (September 1971), 321–39.

LIKERT, RENSIS, *The Human Organization.* New York: McGraw-Hill Book Company, 1967.

MACHIAVELLI, N., *The Prince,* ed. T. G. Gergin. New York: Appleton-Century-Crofts, 1964.

MCCALL, M. W., JR., and M. M. LOMBARDO, eds., *Leadership: Where Else Can We Go?* Durham, N. C.: Duke University Press, 1978.

MCGREGOR, DOUGLAS, *The Human Side of Enterprise.* New York: McGraw-Hill Book Company, 1960.

NEWMAN, WILLIAM, et al., *The Process of Management.* Englewood Cliffs, N.J.: Prentice-Hall, Inc., 1972.

PICKLER, JOSEPH, "Power, Influence and Authority," in *Contemporary Management,* ed. Joseph McGuire. Englewood Cliffs, N.J.: Prentice-Hall, Inc., 1974.

REDDIN, WILLIAM, *Managerial Effectiveness.* New York: McGraw-Hill Book Company, 1970.

STEWART, R., *The Reality of Organizations: A Guide for Managers.* London: The Macmillan Company, 1970, Chaps. 5–8.

STONER, JAMES A. F., *Management.* Englewood Cliffs, N.J.: Prentice-Hall, Inc., 1978.

STUNN, D., "Control: Key to Successful Delegation," *Supervisory Management* (July 1972), 2–8.

TANNENBAUM, ROBERT, and WARREN SCHMIDT, "How to Choose a Leadership Pattern," *Harvard Business Review* (March–April 1958), 95–101.

VROOM, VICTOR H., and PHILIP W. YETTON, *Leadership and Decision-Making.* Pittsburgh: University of Pittsburgh Press, 1973.

WEISSE, PETER D., "What a Chief or Group Executive Cannot Delegate," *Management Review* (May 1975), 4–8.

TECHNIQUES OF ENGINEERING MANAGEMENT

PART III

PART III

DEVELOPING MOTIVATIONAL TECHNIQUES FOR ENGINEERS

The engineer is a professional who is motivated by other than the normal motivators. This requires the engineer manager to understand and identify barriers to individual and team productivity. The engineer manager must develop appropriate techniques for motivating his engineers.

CHAPTER HEADINGS
- ☐ Needs of the Engineer
- ☐ Motivational Characteristics of Engineers
- ☐ Theories of Motivation
- ☐ Applying Motivational Techniques and Concepts
- ☐ Creative Actions That Motivate People
- ☐ Developing Skills
- ☐ Discipline and Morale
- ☐ Summary

LEARNING
OBJECTIVES

☐ Become acquainted with needs of the engineer and learn his motivational characteristics.

☐ Gain knowledge of the various theories of motivation, including those of Maslow, Herzberg, Gellerman, and McGregor.

☐ Learn to apply these motivational techniques to obtain maximum group performance.

☐ Gain insight into developing a team which works toward common goals.

EXECUTIVE COMMENT

HAROLD W. DUCHEK
Senior Vice President, Electronics
and Space Division
Emerson Electric Company

Harold W. Duchek joined Emerson Electric in 1950. He has held positions in design, project/program management, advanced planning, and the internal research and development program. After a period as director of engineering for the Electronics and Space Division, he was promoted to his present position of senior vice president and is responsible for engineering, manufacturing, procurement, and quality assurance.

Thoughts on Motivation

When considering the factors related to the motivation of engineers and technical people in general, two elements evolve as of significant importance: (1) a need to appreciate the unique characteristics associated with this person and his profession and (2) a sensitivity to the fact that as a person he has the same likes, desires, and goals in life, relative to his family, as individuals in any other line of work.

Engineers, and people dedicated to the sciences, usually have probing minds. They have chosen their profession because they like a challenge, and in particular a technical challenge. They seek problems in which their facts are often the fundamentals of the physics books but the application of those facts to the solution of a particular problem may require considerable uniqueness and at times, an invention.

As one considers the motivational factors for individuals in the scientific community, one must recognize the various degrees of that probing mind which one can expect to encounter. On the one extreme there is a constant push for something that is new, the use of the latest materials or the most advanced electronic components, or the application of a recently developed theory in which only the most advanced ideas will be satisfying to the individual. In the middle there is a blend of the new and the practical. Here the term practical is dictated by cost, a schedule, and a predetermined degree of performance. A third level, farther away from scientific newness but more toward the area of total application and the administration of engineering type activities, entails the repetitive application of standard and proven techniques. One needs to recognize the unique desires of the individuals involved and be sensitive to the fact that each will be looking at a particular task through the eyes of what it is he personally wishes to accomplish.

As with any team, if one understands the various participants, their strengths, their weaknesses, and, most importantly, their desires as regards accomplishments, one can make judicious assignments and realize a high degree of motivation. In the practical world, where broad selection in the matching of people types to a requirement is not always possible, an appreciation of people's desires will allow one to compensate for the immediate circumstances through people grouping or personal involvement to assure the desired results.

The second element of importance in achieving motivation in technical people is the maintenance of a sensitivity to the common goals we all have relative to home and family life, regardless of one's state in life. These goals run parallel to the work-oriented technical challenges encountered during the "at work" day. Most individuals keep these two elements separated; however, as one considers motivation there is a need to appreciate the real ingredients of this formula, be able to interpret actions, anticipate problems, and, at times, assist or whatever to maintain adequate emphasis within the job environment.

In summary, then, I believe the common ingredient for motivating engineering personnel is that of understanding: an understanding of what an individual expects in the way of technical challenges; an understanding of how the job-related requirements can be met through a match with the various individuals; and, finally, a recognition of the human side, appreciating that the basics of pay, working conditions, and personal and family goals will strongly influence how an individual perceives a particular assignment. Know, or get to know, your people, for with this understanding motivation so often becomes simple logic.

As an engineer, the engineer manager sought to satisfy his needs. He was motivated to design and develop new equipment, processes, and layouts. His work provided him with many challenges which he sought to meet. He enjoyed attacking problems and overcoming difficult obstacles. In fact, his continued interest was sustained by the degree to which his working experience satisfied his needs. He was motivated by different means than those considered effective for other professional groups.

Now, as an engineer manager, he must be concerned with what needs his people have and how these needs can be satisfied in the work environment. As a manager, he is now in the position to create an environment in which the employee will want to come to work and will get satisfaction from his job.

Motivation might be defined as a force or drive causing some action, behavior, or result. The engineer, like most people, is motivated by certain needs which cause action on his part.

NEEDS OF THE ENGINEER

What are some of the needs of engineers? As a generality, we might say that they are task oriented and internal. Some specific needs are:

1. *Desire to achieve.* The engineer likes to see a goal ahead of him that he can achieve.
2. *Self-expression and creativity.* By nature of his profession, the engineer is constantly creating, designing, and constructing.
3. *Challenge.* Because he is crossing new frontiers, the opportunity of a challenge is always there.
4. *Diversity of problems.* There is little routine in the work of an engineer; once his designs are in operation, new problems arise.
5. *Pride in accomplishment.* The engineer likes to look at his finished product—a bridge, a machine, a plant.
6. *Independence.* The decisions an engineer makes are based on his knowledge and experience. They give a feeling of independence.
7. *Practice of technical knowledge and skills.* He finds pleasure in using his rare talents and skills.
8. *Recognition.* There is a desire on his part to have others recognize his accomplishments.
9. *Professional status.* He desires recognition of his stature in his profession.

If the engineer manager can provide a means of assisting the engineer to satisfy these needs, he has provided a powerful motivating environment. This thought is summarized in Figure 7–1.

FIG. 7–1. Needs–Goals System of the Engineer

*MOTIVATIONAL
CHARACTERISTICS
OF ENGINEERS*
Strong motivators for engineers at work include the opportunity to attack problems directly, to utilize their technical skills and knowledge, and to accomplish project objectives by overcoming difficult obstacles (Table 7–1). These contribute strongly to the satisfaction of the engineer's needs. For the engineer, high satisfaction motivators exist in problem-centered activities; however, to be highly motivated, the engineer must be recognized for his accomplishments. This need for recognition of accomplishments is a need the engineer will have throughout his career.

To be effective, the engineer manager must consider the engineer's role in the organization and how it affects motivation. Some typical roles are listed in Table 7–2.

*THEORIES
OF MOTIVATION*
It is important for the engineer manager to understand the theories of motivation in order to best help his subordinates to meet their needs. The first research into the factors that affect motivation and productivity started in the 1920s with Harvard University's work with the textile industry. The well-known Hawthorne Studies were conducted at the Hawthorne plant of the Western Electric Company in Chicago between the years 1927 and 1933.[1] In these studies an effort was made to relate certain physical conditions, such as lighting, ventilation, temperature, and number and frequency of rest breaks, with work group productivity. A number of surprising results were obtained which utterly failed to relate physical conditions to output. It was found that productivity increased with *any* variance in the physical conditions. It was not dependent on whether physical comfort was increased or decreased. It was not until years later that behavioral scientists were able to unravel the mysteries of these studies. In short, they found that most of the unusual results were due to human and social factors rather than physical. Their interpretations led to an all-out research effort during

---

TABLE 7–1

High Motivators for Engineers

1. Making a direct attack on problems.
2. Being recognized for accomplishments by peers and colleagues.
3. Associating with competent co-workers.
4. Using technical knowledge and skills in making contributions to the advancement of science.
5. Doing new and different things.
6. Being independent.
7. Utilizing professional information to solve problems.
8. Performing job successfully.

---

[1]F. J. Roethlisberger and W. G. Dickson, *Management and the Worker* (Cambridge, Mass.: Harvard University Press, 1939).

World War II to determine how productivity is affected by factors of a human and social nature.

Maslow's Hierarchy of Needs

The next major contribution was the publication in 1954 of Abraham Maslow's basic needs hierarchy, which serves as the source of most of today's motivational theories.[2] Maslow stated that the needs and therefore the wants and drives of a person can be identified by five categories on an ascending scale of influence with a never-ending sequence of needs which motivate him. Maslow suggested that as successive levels of needs are satisfied, other higher-level needs emerge. Maslow suggested the following basic needs (See Figure 7–2):

1. *Survival.* Survival is a universal need which we may define as the need for food, clothing, shelter, and other necessities required to simply maintain life. It is obvious that money is a part of this need, because it is required in order to obtain these necessities. Maslow tells us that survival needs can essentially be fully satisfied. This is the most basic reason why man seeks employment.

2. *Security.* Maslow states that once we have reached a relative level of satisfaction in obtaining those things necessary to maintain life, we are next motivated by the need to keep those things—in other words, security. This need, which is universal among humans, is one which can sometimes be almost totally satisfied in terms of employment. We have all known people whom we considered to be "too security conscious" in their jobs. Symptoms of the security need in managers are fear of risk taking, taking care to cover one's tracks, and management by committee. To unduly threaten a person's security need is to produce safe, conservative, unimaginative results. An insecure person who can be innovative is rare, because all of his actions are designed to preserve the status quo. The security need emerges on a massive scale in times of economic upset and internal corporate stress. Since insecurity almost always results in low morale and low productivity, one of the manager's major tasks is to remove unusual stress from the working climate.

TABLE 7–2

Roles of the Engineer

1. Originates projects; creates; performs research; seeks knowledge.
2. Uses own skills; evaluates own work.
3. Works on limited programs.
4. Has limited responsibilities.
5. Is technically oriented; is specialized.
6. Is objective; is factual.
7. Is independent in action.

---

[2]A. H. Maslow, *Motivation and Personality* (New York: Harper & Row, 1954).

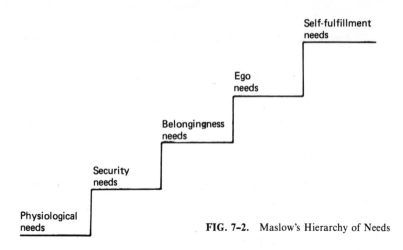

FIG. 7-2. Maslow's Hierarchy of Needs

Labels in figure:
Self-fulfillment needs
Ego needs
Belongingness needs
Security needs
Physiological needs

3. *Belonging*. Belonging is a social need and, according to Maslow, becomes a factor once a relative level of security has been attained. In all phases of human activity, acceptance of the individual by the group is a critical factor in adjustment. If you don't believe it, try living in a community where no one speaks to you or considers you worthy of knowing. In most cases, when an individual cannot become a member of the social group, there is at the very least a productivity problem. Belonging is directly related to the formation of social groupings which tend to have individualistic objectives, patterns of behavior, and codes of rules. Once a social group has formed, the manager is confronted with the task of motivating the individual who is a member of this larger body which controls his behavior on the job. Perhaps here is the beginning of an answer to why financial incentives do not always produce maximum individual output.

These first three basic needs—survival, security, and belonging—represent the so-called lower-level needs which are universal among humans.

4. *Ego*. Maslow describes ego as including the need for recognition of accomplishments and status as denoted by respect from others or by symbols of rank. The need to hold and use power is also a major part of this concept and is often thought to be the prime motivator of persons in politics. The ego need may cause unpredictable behavior in that the smallest slight to a person's ego results in dramatic and long-term effects which seem to go far beyond any possible stimulus. As an example, if the manager is normally cheerful and greets everyone in the office the first thing in the morning, he has established a pattern. See what happens on the rare morning when he is in a rotten mood and goes straight to his office without speaking. Some of his people will take it as a personal slight which reflects on their own worth. Days of sullenness can be the manager's reward. The ego need is both a frustrating factor as well as a strong potential motivator. Unlike the lower-level needs, ego is never really satisfied. As certain levels of satisfaction are reached, new goals are unconsciously set, moving people ever forward.

5. *Self-fulfillment.* Maslow defines self-fulfillment as including challenge, responsibility, variety, and sense of accomplishment. He further suggests that this need is the most difficult to satisfy in many job situations. Like ego, its goals are ever moving and complete satisfaction is never really achieved. An example is a president of the United States who, having achieved fame and wealth, wishes his achievements to attain a significant place in the nation's history. We can easily see that the fulfillment need is closely related to the ego and cannot be separated from it. Maslow feels that it is the most powerful and long-lasting of the motivators. Organizations and managers must be alert in providing for its potential satisfaction in the workplace.

The need categories of Maslow's basic needs model define rather well most of the drives and causes of behavior not only in a working situation but also in all phases of human life. The implications for the manager are that he must have sufficiently close relationships with people to be able to understand and define their outstanding needs. This understanding is an important factor because it is precisely those needs that control human behavior over the long run. Such knowledge will allow the prediction, with some accuracy, of probable reaction to various kinds of management situations. Therefore, the Maslow model becomes a useful tool for the improvement of decision making in the area of individual and group performance.

Herzberg's Studies

Another important contribution to motivation theory was Frederick Herzberg's maintenance-motivator model.[3] In developing this model Herzberg asked 200 engineers and accountants to recall work situation experiences that aroused their emotions. Herzberg and his associates found a two-factor explanation for motivation. His research indicated that one group of factors which were related to the job affected productivity in different ways. He tells us that *increased* satisfaction with these so-called "maintenance" factors, such as company policy, supervision, working conditions, salary, and job security, has little positive effect on performance. On the other hand, *decreasing* satisfaction with these factors has a negative effect on productivity. Therefore, satisfaction with the maintenance factors serves only to maintain morale and attitudes at a given level. They serve only as potential *negative* motivators.

In the second group of factors, Herzberg listed certain satisfiers related to job content. These factors included achievement, recognition, challenging work,

[3]Frederick Herzberg, et al., *The Motivation to Work* (New York: John Wiley, 1959).

responsibility, and advancement. Their existence yields feelings of satisfaction or no satisfaction. They are real motivators because they have the potential of yielding a sense of satisfaction. Motivators are another way of stating Maslow's hierarchy of needs. The dissatisfying factors are monetary and physical in nature and correspond to Maslow's middle plateau of belongingness needs.

Gellerman's Theories

Gellerman has presented some other insights into the use of motivation in management. He speaks of two major styles of management: (1) the method involving coercion, threat of dismissal, and (2) the method involving compensation, reward.[4] Both of these methods assume that people must be manipulated through external controls to be productive. Coercion and manipulation are always with us, but they are no longer considered potent motivators.

Gellerman says the manager must get to the root of man's behavior and must know why a man behaves and acts as he does instead of react to his behavior. This is an analytical approach to understanding people rather than a set of techniques for handling them.

Gellerman further defines morale as the attitude a worker has toward accomplishing his work rather than emotions displayed during work. Therefore, the happiest man is not necessarily the best motivated or the most productive worker.

McGregor's Theory X and Theory Y

McGregor's Theory X and Theory Y are theoretical interpretations of managers' attitudes toward human behavior.[5] Theory X views man as inherently lazy and in need of close control. In this view, man prefers to be told what to do and avoids all responsibility.

Theory Y views man in essentially the opposite way—that man is neither inherently lazy nor inherently productive but that his behavior depends on how he has been treated by his superiors. If he has been closely supervised and given no responsibilities, he will react by being stubborn, uncooperative, or just plain lazy. But if he is given responsibilities and not closely supervised, he will react by being highly motivated and self-controlled and will, in fact, seek more responsibility.

Most people who work for a manager who believes in Theory X will become "maintenance seekers"; most of those who work for a manager who believes in Theory Y will become "motivation seekers." The assumptions the manager makes about people will to a great extent determine whether they are maintenance seekers or motivation seekers.

[4]Saul W. Gellerman, *Motivation and Productivity* (New York: American Management Association, 1963).

[5]Douglas McGregor, *The Human Side of the Enterprise* (New York: McGraw-Hill, 1963).

Texas Instruments Studies

About two years after Herzberg published his findings, a group of research psychologists at Texas Instruments decided to test Herzberg's theories on their own people.[6] They followed the same interview pattern used by Herzberg, but they expanded the groups interviewed to include scientists, engineers, manufacturing supervisors, hourly male technicians, and hourly female assemblers. Their study supported Herzberg's conclusions that motivation stems from the challenge of the job through such factors as achievement, growth, responsibility, work itself, and earned recognition. These motivational needs are shown in the inner circle of Figure 7–3. Satisfaction of these needs can be accomplished by the mechanisms and factors listed in the inner circle.

Skinner's Operant Conditioning

B. F. Skinner, a Harvard psychologist, has a theory known as *operant conditioning,* referring to his standard practice of making a subject operate in a certain way to receive a certain reward.[7] His idea is that man is what he is because of his environment, not because of any internal drives, needs, or other unexplainable influences. He defines an operant response as the desired behavior probably followed by a pleasant experience. Since we prefer pleasure to pain, the subject repeats that which brings pleasure.

Jablonsky and DeVries have suggested several ways for achieving motivation through operant conditioning. A few of them are:

1. Avoid relying on punishment as a primary means of motivation.
2. Positively reinforce the desired behavior and if possible ignore the undesired behavior.
3. Minimize the time lag between the operant response and the reinforcement.
4. Determine the environmental factors that are considered positive and negative by the individual.[8]

Vroom's Preference-Expectation Theory

In 1964 Victor H. Vroom described how preference and expectation work on each other to determine motivation.[9] Preference refers to the multiple possible outcomes that a worker might have for an activity. He has a preference for certain rewards related to his performance. The worker also has a certain expectation that the desired outcome will happen. His enthusiasm to receive a reward is tempered by his expectation.

[6]M. Scott Myers, "Who Are Your Motivated Workers?" *Harvard Business Review* (January–February 1964), 73–78.

[7]B. F. Skinner, *Service and Human Behavior* (New York: Macmillan, 1953).

[8]S. Jablonsky and D. DeVries, "Operant Conditioning Principles Extrapolated to the Theory of Management," *Organizational Behavior and Human Performance* (April 1972), 340–58.

[9]Victor H. Vroom, *Work and Motivation* (New York: John Wiley, 1964).

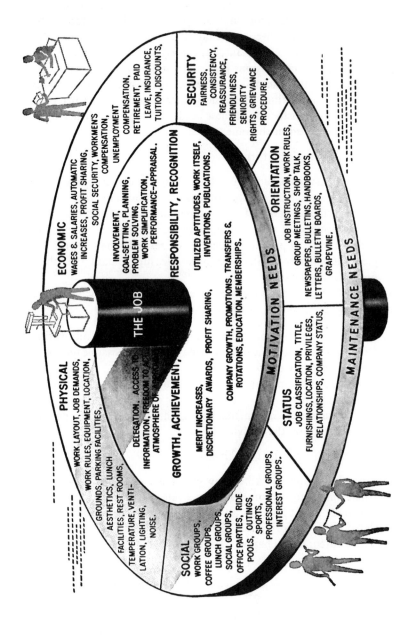

FIG. 7-3. Employee Needs–Maintenance and Motivation. [Reprinted by permission of the Harvard Business Review. Exhibit from "Who Are Your Motivated Workers?" by M. Scott Myers (January–February 1964). Copyright © 1964 by the President and Fellows of Harvard College; all rights reserved.]

Porter and Lawler's Model

Porter and Lawler expanded Vroom's theory into the concept that satisfaction is determined by how clearly the actual rewards given compare with what the worker feels he deserves.[10, 11] If the reward equals or exceeds his expectations, the worker will be motivated to repeat his actions. If the reward falls short of his expectations, the worker's dissatisfaction will prevent him from being motivated to continue his effort.

Frustration and Unsatisfied Needs

Frustration is the feeling of insecurity and dissatisfaction that arises from unresolved problems and unsatisfied wants.[12] Through a type of behavior known as *adjustive reaction,* the mind generally attempts to cause behavior designed to aid the frustrated person to adjust to an unresolved situation. Some adjustive reactions are positively directed and others are negatively directed. An understanding of psychological concepts such as these enables us to be sensitive to our own complex mental problems and those of others. An awareness of major adjustive reactions may enable us to deal more effectively with the normal stresses of everyday living.

APPLYING MOTIVATIONAL TECHNIQUES AND CONCEPTS

Since the engineer manager works with people, learning what motivates them is important. He needs to understand why they behave as they do. This understanding is necessary to induce his people to effectively do the work they have been trained to perform. His objective should be to provide as many motivators in the workplace as possible, rather than forcing his subordinates to look for their motivators outside the workplace. People will always receive some motivation from outside work activities, but the engineer manager can have a considerable influence on even these motivators. By knowing his employees' needs and motivators, he is able to predict what will happen when a decision involves his subordinates. It is important, therefore, for him to know motivation concepts, since these can be useful in a number of ways.

Using the motivation theories of the last section as guides, let us examine more specific ways that the engineer manager may motivate his subordinates. Important motivators include:

1. *Salary and wages.* These are generally based on the firm's wage policies and on competitive practices. The engineer manager has only limited authority over salaries and wages, and his salary recommendations must be made within a narrow range. Fringe benefits provide added incentives to subordinates but are generally beyond his control. Therefore, the effect of these areas as motivators is controlled largely by company policies. By denying salary increases, however, the manager may provide a negative motivator or perhaps spur the employee to improve his performance.

[10]Edward Lawler and L. Porter, "The Effects of Performance on Job Satisfaction," *Industrial Relations* (October 1967), 23.

[11]L. Porter and Edward Lawler, *Managerial Attitudes and Performance* (Homewood, Ill.: Richard D. Irwin, 1968).

[12]S. Kessen, *The Human Side of Organization* (New York: Harper & Row, 1978), pp. 117–26.

2. *Job security*. This assures the fulfillment of material needs over a period of time. The engineer manager has considerable influence over the degree of security that a subordinate feels in his work; e.g., by assuring the subordinate of his satisfactory performance, by salary recommendations, and by promotions.

3. *Status*. This reflects the desire of the engineer manager's subordinates to have recognition, both inside and outside the organization. People demand or want society's approval of their work and efforts. This motivational need can be satisfied through promotions and status symbols. Promotion is a reward for the individual's ability and is a very effective motivational tool. Other examples of status symbols are a reserved parking space identified with the individual's name, the job title, enclosed office, new office equipment and furniture, and secretaries. To be effective, the engineer manager must carefully use these techniques as rewards for outstanding performance.

4. *Social relationships*. These are among the great benefits derived from working. When the engineer manager is able to develop an environment in which the individual is provided social relationships by exchanging news, ideas, confidences, and companionship, the workplace and co-workers fill a very important individual need. If provided in the workplace rather than outside, there is less desire to get away to fill these needs. Social relationships can be enhanced through stability of the work force, carefully considering transfers, and other actions.

5. *Power*. This is an indicator of the need for respect from others. To be able to hold and use power is a major part of this motivator. Power is a prime motivator for some engineers, since their desire to be respected and acknowledged is important. As with other motivators, such a need is never really satisfied and as certain levels of satisfaction are reached, new power goals are unconsciously set.

6. *Participation*. As a motivator, participation gives the engineer the opportunity to feel that he possesses valuable information that will improve the productivity and efficiency of the organization. In many situations, the engineer may have valuable insight into methods of solving a problem. Accepting and using such views, expertise, and ideas are part of the motivational process. Some contributions may not be relevant, but the employee feels he has at least had his chance to be heard. Although the engineer manager cannot request the views of his people on all problems, he should occasionally use the participation approach.

7. *Creativity*. This is that quality required to produce new and practical ideas or productive innovations. Most engineers are highly motivated to solve difficult problems and to see the solutions put into operation. The engineer manager must make an attempt to recognize and encourage these individuals.

The engineer manager can use these techniques to good advantage. Although the motivational techniques available are many, the engineer manager must know his people if he is to use them effectively.

In practice, the engineer manager must take several combined actions to motivate an individual. No single set of actions will work for everyone. The engineer manager should keep the following concepts in mind:

1. People must know where and how they fit into the accomplishment of goals or objectives. The engineer manager cannot assume that because an employee has been around, he automatically knows the operations or can identify his role. It is the engineer manager's responsibility to make their roles clear and meaningful to employees.

2. The engineer manager must always promote and assist his subordinates' advancement in the organization. This involves the following roles for the engineer manager:

    A. Training subordinates to accept both technical and managerial responsibilities. They thus become more valuable to the department and the organization.

    B. Making subordinates aware of opportunities in the organization and assisting them in obtaining advancement. He must give his subordinates every opportunity to improve.

These roles may appear to be counterproductive in that the engineer manager may appear to be training people for someone else in the organization or even for other firms. In the long run, however, his organization will have the reputation of being an excellent place to train for promotion. This will attract outstanding people and result in highly motivated employees who have great enthusiasm for their work.

3. The action in an organized group is out on the production line or project, not in the engineer manager's office. Therefore, it is absolutely necessary for him to get out of his office to know what is going on. Otherwise, he will base his decisions on second-hand information from his staff and secretary rather than on information from people on the job. The engineer manager must get out and meet employees and visit with them as they work. This takes effort on the part of a very busy individual. If he goes into an area and the supervisor is not around, a note letting the supervisor know of his visit is necessary. This shows that an effort has been made to see both the supervisor and his subordinates. After a few visits, the engineer manager will be surprised how much he has learned and how important these visits are to him. He must make such visits a habit if he is to be a successful manager.

4. A very effective management technique is to send a note to the family when an employee accomplishes a project on time, works overtime during evenings or weekends, or engages in other events affecting his family. The manager should state how much he appreciates this effort and recognizes that the family has made a sacrifice. Also, inviting the family into the plant and letting the employee show the family around may be very important to the employee and his family.

5. Delegation of activities to be performed by others makes work more meaningful. Delegating shows respect for an employee's judgment and ability, fosters his initiative and confidence, and makes him feel more important. The engineer manager must make positive efforts to encourage his employees to accept the practice of delegation.

6. Participation in decisions and actions affecting their work or workplace plays a large part in gaining commitments from employees. When one participates in a decision, he becomes mentally and emotionally as well as physically involved. The subordinate's role is now active rather than passive. Participation involves actively seeking ideas and suggestions and encouraging subordinates to voice their opinions and to get involved, thus allowing them to contribute their skills.

Major changes proposed without employee involvement result in greater opposition and less success in implementation. The engineer manager often finds that changes can be made with hearty support and enthusiasm from employees if they are asked for their opinions. Otherwise, the consequences can have a major negative effect on morale and perhaps be expensive.

7. Communication is the key to developing better relationships with people. Most engineer managers are not conscious of how much adequate communication motivates people, especially those who are technically trained. By communicating effectively, he will discover how responsive people can be.

8. Job enrichment is a technique used to build into the job a higher sense of challenge and achievement. A job may be enriched by giving it variety. It may also be enriched by (1) giving the worker more latitude in deciding about such things as work methods, (2) encouraging participation of subordinates and interaction between workers, (3) giving workers a feeling of personal responsibility for their tasks, and (4) taking steps to make sure people can see how their tasks contribute to a finished product and the welfare of the enterprise.

DEVELOPING SKILLS

Influencing others is perhaps one of the principal and most important of the engineer manager's activities. He can better develop his skills and his ability to influence the behavior of others by studying the results of his actions. For example, in a meeting the engineer manager may feel the group is not making progress and is becoming bored with all the unnecessary and unrelated talk. He may enter the discussion and make some strong suggestions about what should be done. The others may look at him with irritation or even mutter comments he cannot hear. So he concludes that either he "blew it" or the group is too stupid to waste time with.

In either case, the engineer manager has done nothing to develop his team-building skills. He has perpetuated his ineffective skills and, because he never questioned what went wrong, he assumed his approach was correct. He should have evaluated his action from the standpoint of the following questions: Was the discussion really important? Was his suggestion appropriate or abrasive? In this manner, he is not blaming anyone; instead, he is reassessing his skill in influencing others.

The engineer manager must evaluate his behavior as others see it, not as he sees it. If he is to develop skills for influencing the responses of others, he must discover how his behavior appears to others. This requires learning to correctly read the continuous feedback provided by the group or learning what behavior creates unwanted responses. With this skill, the engineer manager can sense the conditions and modify his behavior. Then he must embark on the discovery of more effective techniques. He must learn to read interpersonal situations more accurately and improve his effectiveness when interacting with others.

One effective way to discover weak or unresponsive behavior skills is to ask people he knows and respects. Most people who seem to be particularly knowledgeable in interpersonal behavior and able to sense the group's feelings will be eager to assist the engineer manager. Also, he can observe the behavioral skills of those who are effective and try their techniques.

The basic skills needed to change a group of individuals into an effective team are (1) the ability to understand team characteristics and (2) the ability to manage the team.

Ability to Understand Team Characteristics

The process of understanding the team should begin with an understanding of the team's strong and weak points. Frequent evaluation is necessary, for changes in experience and skills acquired by the group are constantly occurring. Such information obtained through continuous feedback can identify areas of performance where knowledge or skills are weak. Understanding the team's characteristics is the engineer manager's responsibility, because he is trying to influence the perception, feeling, and behavior of the group. The feedback process, if the engineer manager learns to use it, will provide information on how effective he is.

To analyze the team characteristics, the engineer manager might ask the following questions:

1. How does the team generally interact with or relate to other groups in work or social situations?

2. Are there any initiators who try to get others to join in doing things? Or do some members of the group usually wait for others to invite them to join in?

3. Do members of the group express warmth and friendliness toward others?

4. What things do they do well?

5. What has the group got going for it?

6. What are the team's goals and does the team have a clear understanding of how to achieve these goals?

7. How well does the team organize its work?

8. What are the team's major weaknesses?

9. What are the distinctive motivational patterns, individual self-concepts, and unique prior experiences of the group members?

10. What are the values and personal goals of the team members? What is important to team members? What do team members hope to get out of participating? Companionship, recognition, skills, acceptance and friendship, experience?

11. What is the team's behavior toward dominant needs and what is the team doing to satisfy them? What seems to turn the team off?

12. What past experience has the group had? Similar activities? What was its role? What special skills do the team members have?

13. What interpersonal relationships exist?

14. Do team members understand their roles, plans, and responsibilities?

The key to understanding the team lies in learning the team's characteristics and interpreting the feedback information. This can improve the manager's effectiveness and help him to adjust his behavior. Understanding the team is a very important part of his role as a team leader. He should understand the team's activities, attitudes, motivational patterns, prior experience, and other characteristics, for he must evaluate and relate them to future tasks the team will be assigned.

Ability to Manage the Team

Another important aspect of developing an effective team is the ability to manage the team. This requires the engineer manager to acquire team leader skills. Acquisition of these skills comes from using his own experience and carefully following the rules for team management. Some important tools for the manager are the following:

1. *Planning team performance.* Superior team performance is the result of careful planning by the engineer manager. For most team members, planning can be very frustrating; therefore, the engineer manager needs to present a broad tentative plan to the team members in order to obtain the benefits of team involvement. By this method, team members will have an opportunity to become involved in the plan's details. Otherwise, their efforts may be fragmented, as illustrated in Figure 7–4. Some guidelines for developing a plan are:

A. *Developing goals for the team.* These are statements on how the team will reach success and statements on what is to be achieved within a specific time frame. When the engineer manager defines the goals in specific terms, the planning process is more realistic. It is his responsibility to make these goals realistic to the team. "Reduce costs" does not tell the subordinate much, but "reduce labor cost by 5%" is very meaningful, particularly if the subordinate participated in making the decision. All goals should be a challenge, ones the team will need to work to achieve. The engineer manager must make certain that he does not expect the impossible, and the team must feel the goals are achievable.

B. *Planning.* This requires defining all tasks that must be accomplished to achieve the goals set forth for the team. The engineer manager can play a key role in assisting the team to define the various functions that must be accomplished. This should involve team participation because the engineer manager may overlook some function to be performed. Participation provides an opportunity for the team members to know what is expected of them in reaching the goals.

C. *Establishing standards.* Because all tasks are interdependent and influence the overall team effort, standards must be developed by the engineer manager. If these prove impossible to meet, corrective action must be taken through redistribution of work tasks.

D. *Feedback controls.* The engineer manager must be provided information on the progress of various tasks and of the team as a whole. Effective feedback helps to provide a cohesive team and is a good motivator. It also provides individual team members with assurance that everyone is pulling together toward common objectives.

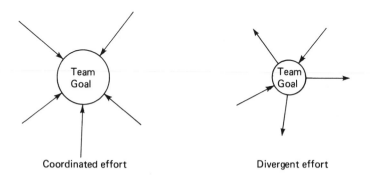

Coordinated effort   Divergent effort

Team Member Actions

FIG. 7-4. Effect of Divergent Team Members' Efforts

2. *Team communication.* A communication system is necessary to achieve an effective environment for two-way flow of information between the engineer manager and team members, as well as between members. An effective communication environment requires the engineer manager to:

A. Demonstrate fairness and integrity with team members.

B. Provide team members complete and timely information, rather than letting them find out news through others.

C. Assist and work with team members by helping them with their tasks and discouraging actions that would cause insecurity, complaining, and defensiveness.

D. Develop a relationship with team members so they feel he is part of the group.

E. Consider the ideas and views of all team members regardless if they differ from his own and convey a feeling of respect for their ideas.

These practices will provide the team with an effective communication system. The engineer manager, however, must still get his ideas across, not only to team members but also to many others outside the team.

3. *Team organizing.* The engineer manager must organize available team resources to accomplish the tasks to be done. Through planning, the various tasks are defined so that each team member has a clear idea of what is expected of him and how his efforts will be coordinated to meet the team goals. The engineer manager must make sure that these tasks are assigned in such a fashion that team members feel that the workload is distributed fairly. Necessary resources provided must include counseling on how to accomplish these tasks.

4. *Team development.* An effective, cohesive, highly productive team requires the engineer manager to develop high morale within the group, interaction among team members, and growth of individual team members. He must always be concerned with how the team members' activities interrelate. The level of team interaction is an important factor in accomplishing team goals.

One of the most effective ways to develop and improve team interaction is through team meetings. During the team meeting the engineer manager can greatly influence the degree of interaction of the team members. This requires him to make team meetings productive and to minimize spurious inputs from team members. Team members' energies are to be directed to constructive activities rather than to destructive attacks on fellow members. One effective method is for the engineer manager to encourage participation and himself assume an active role. To improve the team's interaction, the engineer manager can practice the following:

A. *Make suggestions* for solutions to problems and introduce new ideas.

B. *Ask questions* to ensure that all team members understand what is being discussed and offer comments to assist members to understand and clarify their ideas.

C. *Analyze,* restate, and summarize the points that have been made by the team so that the discussion concerns the question at hand and avoids wasting team time. Bringing pertinent points together enables each team member to determine his own level of understanding up to this point of the discussion. Such statements as "No one seems to like this approach" or "Everyone agrees that we should follow the ABC approach" provide an opportunity for those who *disagree* to express their feelings.

D. *Look* for an expression of feeling from the team members. These expressions may be verbal or visual as indicated by lack of interest or other physical activity. The engineer manager gains information on where the team members stand on the subject. Constant surveillance is necessary during the discussion to determine whether or not there is enough support to continue discussing the question or whether or not ideas must be obtained from team members who are not participating.

E. *Inform* the team when decisions conflict with company policies. This will prevent having to remake decisions at a later time.

F. *Evaluate* team activity during the meeting in the light of team goals and objectives. The engineer manager has the responsibility to point out decisions that are weak and will not achieve the team's goals. He must do this in a way that does not put team members on the defensive but instead reinforces what has been accomplished so they will continue working on the problem. This evaluation by the engineer manager will reinforce the idea that all team members are responsible for team decisions.

G. *Determine* the cause of conflict among team members making decisions and develop appropriate steps to overcome this conflict. This also requires the engineer manager to examine interaction among team members.

H. *Negotiate* compromise solutions that will be acceptable to the disagreeing members of the team. The engineer manager must prepare to accept compromise. For each problem brought before the team, he should have a compromise solution for the team to consider rather than requiring the team to develop one. He must be careful to ensure that differences are actually reconciled and that compromises are accepted rather than forced on the team.

By following these principles in team meetings, the engineer manager will help the team members arrive promptly at decisions in minimal time and with an acceptable degree of conflict. By providing such leadership, the engineer manager will help the team achieve the group relationship and interaction necessary to be effective.

5. *Avoiding team conflicts.* The engineer manager must strive to minimize team conflicts. If he senses such conflicts and is prepared to deal with them immediately, in most cases, he will be able to avoid harm to the team members. Some of the most common team conflicts arise from:

A. Criticizing or blaming others
B. Bringing up ideas and examples unrelated to the subject
C. Arguing for the sake of arguing
D. Loud or excessive talking
E. Talking or carrying out other activities when others have the floor
F. Disrupting the work of the team
G. Introducing suggestions that are irrelevant or lack support

When individual team members follow such practices, action should be taken to correct such behavior by helping them to develop as productive team members. Unfortunately, in most cases, the engineer manager either ignores them or tries to remove them from the team.

Developing skill in team building is one of the most exciting and one of the most difficult challenges faced by the engineer manager. Most people are required to work hard to develop these skills. Development of team performance is the foundation on which all successful team building is based. Effective team performance comes from creating an environment in which the team members work together in a single group.

DISCIPLINE AND MORALE

One of the most difficult tasks for the engineer manager is disciplining his subordinates. All organizations have certain rules and policies that all members are expected to follow. In order to enforce these, discipline is used with the intent of training employees to behave according to the rules and policies. Most engineer managers avoid this type of action, because they realize that to punish a person may cause him to conceal the criticized behavior from the manager in the future. They would rather correct unsuitable behavior by withholding some part of an effective reward system rather than by punishment. This requires the engineer manager to determine what reward(s) can eliminate the unsuitable behavior and bring about the desired conformance. This is a more indirect form of discipline.

Disciplinary standards are established because such outside sources as a union or government have regulations requiring the use of a formal system to prevent unfair discipline. Therefore, some organizations have developed very detailed policies in which written notes must be given the employee for layoff, pay reduction, or other disciplinary actions.

One important action the engineer manager can perform is to use both verbal and written communication in alerting employees to disciplinary practices (Figure 7–5). He must make sure all understand them and understand the reasons for such rules. Also, it is necessary to treat everyone the same and not discipline one person severely and another lightly for the same offense. Such inconsistency

EMPLOYEE/MANAGER	EMPLOYEE RECORD
Informal discussion and policy restatement	Note of discussion placed in file
	Record of conference in file
Oral warning conference	
	Copy placed in file
Written warning	
	Appropriate information placed in file
Discharge	

FIG. 7-5. Disciplinary Procedure Commonly Followed

interferes with the corrective effect of discipline and soon the engineer manager will find that he has lost his effectiveness.

Because disciplinary actions in an organization are very common, it seems impossible to have an organization without them. Therefore, the engineer manager must try to make discipline a form of training and growth. This requires mutual understanding, confidence, and a close working relationship with his subordinates. When such actions are properly used by the engineer manager, the subordinate may gain a favorable attitude and there may even be an uplifting effect on the subordinate. Some principles on handling disciplinary matters are given below:

1. The goals and objectives of the organization must be explained in meaningful terms so that each team member knows why his tasks are so important and that others are depending on him.

2. A reasonable number of disciplinary rules should be developed that can be understood by all people. They should be made available to and explained to subordinates.

3. Disciplinary rules should be brought to the attention of employees from time to time.

4. Adequate written records should be maintained of all disciplinary actions.

5. Uniform discipline should be followed for all members of the group.

6. The real purpose of discipline is to prevent future offenses and to direct and motivate the employee in the desired direction.

7. Disciplinary rules should provide an incentive to correct the fault.

8. Discipline merely for the sake of showing authority is ineffective and results in only minimal improvement in performance.

9. Continual review and update of disciplinary rules and elimination of those no longer applicable are necessary.

10. Use assertiveness as necessary.[13]

Discipline helps the engineer manager ensure that his people will work for the good of all. For the organization to function, reasonable discipline must be maintained as a means of motivating people. Discipline still has the meaning of punishment to some people. This, however, is the last resort of effective discipline and should be seldom used by the engineer manager. To overcome such attitudes, all discipline rules should include (1) acceptable performance, (2) how to achieve such performance, and (3) results of satisfactory performance. By making sure each discipline rule has these three characteristics, the engineer manager will be certain that each one is reasonable and therefore will be effective. Of greater significance is the acceptance of discipline rules by the group. Acceptance can be achieved by seeking their participation in developing acceptable performance standards.

SUMMARY   Motivation can be defined as a force or drive causing some action, behavior, or result. In the case of the engineer, the needs that motivate him are somewhat different from those of other individuals and include desire to achieve, challenge, diversity of problems, independence, recognition, and professional status.

Knowledge of the theories of motivation is important to the engineer. Work in this area began with the Hawthorne studies in the 1920s. They were followed by the development of Maslow's hierarchy of needs which includes survival, security, belonging, ego, and self-fulfillment. Herzberg expanded on these and found that certain maintenance factors had little positive influence but they could have significant negative influence. McGregor developed Theories X and Y which represent theoretical interpretations of managers' attitudes toward human behavior. Theory X views man as inherently lazy and in need of close control. Theory Y views man's behavior as depending on how he is treated by his superiors.

Texas Instruments expanded on Herzberg's studies, and several other theories have recently been developed, including Skinner's operant conditioning, Vroom's preference–expectation theory, and Porter and Lawler's model. These various motivational theories can be applied to such matters as salary and wages, job security, status, social relationships, power, participation, and creativity in order to provide motivation to the engineer.

The manager may take a number of actions in order to motivate individuals, including identifying for them where they fit into the accomplishment of goals or objectives, providing for their advancement in the organization, spend-

[13]Manuel J. Smith, *When I Say No, I Feel Guilty* (New York: Dial Press, 1975).

ing time with them in the field, using delegation to give them meaningful work, providing adequate communication, and providing job enrichment through giving them a higher sense of challenge and achievement.

The engineer manager must develop an ability to understand the characteristics of his team. His ability to manage his team is enhanced through planning for team performance, providing adequate communication, organizing the team effectively, and developing it into an effective group whose morale is high.

The manager must effectively use meetings with his team to weld it into an effective unit. He can do this through making suggestions, asking questions, analyzing their feelings, informing them of company policies, evaluating their activities, and negotiating compromise solutions when necessary.

Even though the engineer manager hates to enforce discipline, it is necessary that he do so for the long-term well-being of the team. If properly applied, discipline can improve the morale of the overall unit.

Important Terms

Discipline:	Correction of unsuitable behavior in order to encourage employees to meet established standards and rules in the future.
Job enrichment:	A technique used to build into a job a higher sense of challenge and achievement.
McGregor's Theories X and Y:	Theory X views man as inherently lazy and in need of close control. Theory Y views man and his behavior as depending on how he is treated by his superiors.
Maslow's Hierarchy of Needs:	The five basic needs reaching from the lowest to the highest levels of satisfaction are survival, security, belonging, ego, and self-fulfillment.
Motivation:	The force or drive which causes some action, behavior, or result.

For Discussion

1. What qualifications would you look for in a team leader? What kind of personality, experience, ability to organize, knowledge, etc., is needed?
2. Describe Maslow's basic needs hierarchy and explain how each item affects employee behavior and productivity.

3. Explain how participation can be used as an important motivational technique.
4. Compare and constrast Maslow's theory with Herzberg's theory.
5. Discipline is an important aspect of managing. What are some effective means of achieving discipline?
6. Compare the motivational factors of a production line employee and a professional employee.
7. What two basic skills are needed to develop a group of individuals into a team? How can these be implemented?
8. Why is a sense of belonging important for the employee and how can the engineer manager help the employee feel that he belongs?
9. Why do engineer managers tend to dislike the direct disciplining of subordinates? What other methods may he use?
10. How could an engineer manager best motivate his engineer subordinates?
11. What important skills should an engineer manager develop in order to manage a team effectively?
12. How do you develop and manage an effective team?
13. Once the engineer manager finds out what motivates a particular individual, can he use that motivator to motivate all his employees? Comment and discuss.
14. What can the engineer manager do to improve member interaction at team meetings?
15. What are some of the causes of team conflicts and what can be done to prevent and correct these conflicts?

Case 7-1

What Motivates the Engineers?

Rex feels that the engineers are not motivated toward superior performance. Without any specific plans to change the situation, he had the following visit with the firm's personnel manager:

Rex: "I just can't seem to get my engineers to perform. They are all extremely competent, but they just aren't willing to exert the extra effort required and expected if R&D is going to help our firm to be a leader in the industry. As you know, they perform some important functions in new product design, modification of existing products to meet competition, and help our customers use our product."

Personnel Manager: "How do you evaluate their performance?"

Rex: "Mainly on whether or not they meet project deadlines. It is hard to evaluate the quality of their work since it is always different, and designs are frequently altered later to facilitate production or meet changing customer requirements."

Personnel Manager: "Are they meeting deadlines regularly?"

Rex: "No, and they don't really seem too concerned about it."

Personnel Manager: "If they meet a deadline, is there any financial reward?"

Rex: "No, not really. As you know, company policy doesn't allow us to provide any extra financial compensation during the year. About the only tool I have is when I pass out the yearly raise money. Of course, they don't have any basic salary complaint because we are above average in our industry and our profit sharing plan is among the best."

Personnel Manager: "How about promotions?"

Rex: "We have several different grades within R&D."

Personnel Manager: "Have you considered dismissing any of them?"

Rex: "No way! We need them too much and it is impossible to replace them, to say nothing of being too expensive. Also, Tom Johnson would not allow this because we are so far behind in our work."

1. What is causing this apparent lack of motivation?
2. What should the personnel manager suggest to Rex?

Case 7–2

Surprised Motivation

After Rex Smith had been chief engineer for a couple of years he became somewhat unhappy with his engineers' performance. For one thing, they didn't get their reports in on time. No matter how much he talked to them, coached them, or wrote memos to them, they didn't seem to appreciate the need for reports. As a result, no improvement followed. He finally came to the conclusion that he knew of no way to make them change their performance.

There had been some work underway for a year or so on building a new office building. The engineering department was to be headquartered in this new building. The building had been well designed and there were private offices for the project managers and a large amount of new office equipment for dictating and typing. There were also some small offices where engineers could work privately in developing reports or doing library-type research. All in all, it seemed to Rex a nice installation.

Some time after the department had moved into the new building, Rex was reviewing the progress of a number of projects when to his surprise he noted that reports seemed to be coming in on time. He was dumbfounded. He hadn't said anything about reports not being on time since the department moved into the new building. He hadn't even written a memo on it.

1. What could have brought about the behavior change?
2. Of the possibilities, which do you think is most significant?
3. What action should Rex take to keep the reports flowing in on time?

For Further Reading

ATKINSON, J. W., and J. O. RAYNOR, *Motivation and Achievement*. New York: Holt, Rinehart & Winston, 1974.

BAYTON, JAMES A., and RICHARD L. CHAPMAN, *Transformation of Scientists and Engineers into Managers*. Washington, D.C.: NASA SP 291, 1972.

COOPER, R., *Job Motivation and Job Design*. London: Institute of Personnel Management, 1974.

DAVIS, K., ed., *Organization Behavior: A Book of Readings,* 4th ed. New York: McGraw-Hill Book Company, 1974.

GELLERMAN, SAUL W., *Motivation and Productivity*. New York: AMACOM, 1963.

HACKMAN, J. R., et al., "A New Strategy for Job Enrichment," *California Management Review* (Summer 1975), 57–71.

HERZBERG, FREDERICK, *The Motivation to Work*. New York: John Wiley & Sons, Inc., 1959.

KORMAN, ABRAHAM K., *The Psychology of Motivation*. Englewood Cliffs, N.J.: Prentice-Hall, Inc., 1974.

LAWLER, EDWARD E., *Motivation in Work Organizations*. Monterey, Calif.: Brooks-Cole, 1973.

LOCKE, EDWIN A., "Personnel Attitudes and Motivation," *Annual Review of Psychology* (1975), 457–80.

MAHLER, W. R., and W. F. WRIGHTNOUR, *Executive Continuity: How to Build and Retain an Effective Management Team*. Homewood, Ill.: Dow-Jones-Irwin, 1973.

MASLOW, A. H., *Motivation and Personality*. New York: Harper & Row Publishers, Inc., 1954.

MCCELLAND, DAVID C., *The Personality*. New York: The Dryden Press, 1951.

PENNINGS, J. M., "Work Value Systems of White-Collar Workers," *Administrative Science Quarterly,* 15, No. 3 (1970), 397–405.

SCHNEIER, CRAIG, "Behavior Modification in Management: A Review and Critique," *Academy of Management Journal* (September 1974), 528–48.

SKINNER, B. F., *Science and Behavior*. New York: Macmillan, Inc., 1953.

STEERS, RICHARD, and LYMAN PORTER, *Motivation and Work Behavior*. New York: McGraw-Hill Book Company, 1975.

STEIN, BARRY A., et al., "Flextime: Work When You Want To," *Psychology Today* (June 1976), 40.

"Team Spirit Results in Higher Productivity, Job Satisfaction," *Commerce Today* (September 30, 1974), 5.

VROOM, VICTOR, *Work and Motivation*. New York: John Wiley & Sons, Inc., 1964.

WEINER, BERNARD, *Theories of Motivation*. Chicago: Markham Publishing, 1972.

MAKING EFFECTIVE APPRAISAL EVALUATIONS AND PERSONAL DEVELOPMENT

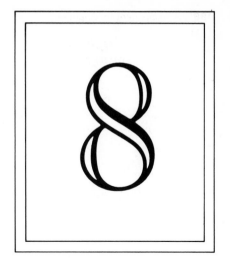

The engineer is knowledgeable in appraising machines and developing new technology. Now it is critical for the engineer manager to effectively appraise people and develop people to their maximum potential as engineers. Such factors as uncertainty of outcome, lack of needed information, unknown factors, and subjective judgment cause particular difficulty for the engineer manager.

CHAPTER HEADINGS
- ☐ Problems in Appraising Engineers
- ☐ The Appraisal Process
- ☐ Approaches to Appraisal
- ☐ Training and Development
- ☐ Training Programs
- ☐ Personal Development Programs for Engineers
- ☐ Organizational Development
- ☐ Summary

LEARNING
OBJECTIVES

- ☐ Gain insight into the importance of the performance appraisal.
- ☐ Learn the various approaches to the appraisal process and their advantages and disadvantages.
- ☐ Understand the method for conducting the appraisal.
- ☐ Learn the various methods for training subordinates in preparation for future promotion.

EXECUTIVE COMMENT

MERTON G. WALKER, JR.
Corporate Director — Technical Administration
McDonnell Douglas Corporation

Merton G. Walker joined McDonnell Douglas Corporation in 1946 as a design engineer. Through the years he advanced through various engineering and managerial positions to his present position of corporate director of Technical Administration. He is the author of numerous articles on engineering administration and engineering personnel.

The Benefits of the Performance Appraisal Process

Appraising employee performance is an essential part of every manager's job. The effective utilization of human resources requires a manager to know each individual employee's personal goals, needs, likes and dislikes, weaknesses, strengths, etc., in order to apply his talents most appropriately. This in turn requires conscientious appraisals tailored to organizational and individual goals; discussion, agreement, and commitment regarding job responsibilities and expected results; evaluation and communication of employee performance; and documentation of the appraisal.

Two generally accepted appraisal techniques are interviews and job evaluations. Each technique has advantages and disadvantages, but when used together, they complement each other very well and produce outstanding results.

The written performance appraisal has several benefits:

1. It gives the manager a convenient worksheet for organizing his appraisal and ensuring its completeness.
2. It allows the manager's opinions to be reviewed and approved by the next level of supervision.
3. It constitutes a valuable record for future salary and promotion reviews.
4. It forms a basis for the organization's defense against potential charges of discrimination in connection with raises, promotions, and discharges.
5. It can be used as a checklist and guide during the oral interview.
6. It can be used to record the results of the interview.
7. It can be given to the employee to document his commitment to stated goals.

The written appraisal is incomplete by itself. It is a form of one-way communication, and it therefore tends to keep the manager in the dark about his employees' feelings and aspirations. It must be reinforced by a personal appraisal interview with each individual employee.

The appraisal interview has many potential benefits:

1. It affords a chance for two-way communication about matters of vital concern to both the manager and the employee.
2. It allows the manager an opportunity for coaching, counseling, and career guidance.
3. It can be used to motivate the employee to higher standards of performance.
4. It involves the employee in determining plans, goals, and commitments concerning job performance. Extensive studies have shown that people tend to be highly committed to plans and goals that they have helped to establish; they become ego-involved.

Unfortunately, some engineer managers shy away from face-to-face discussions about an employee's job performance, areas that need improvement, and career growth opportunities. They find that such confrontations make them feel uncomfortable. My experience tells me (1) that such people do not fully understand their role in the performance-appraisal process (they are helpful counselors and advisers, not critics) and (2) that they have not been properly trained for their managerial function.

Engineers are constantly evaluating the work they are involved in and are always trying to apply new technology to make the equipment or machine operate more efficiently, but these evaluations are always in physical terms or units. These evaluations are always in comparison to some standard or theoretical relationship. In a production line the engineer attempts to have the equipment operate at a given level as set by the manufacturer or to do test runs to determine the machine's maximum capacity. In other problems, such as maintenance, the engineer tries to develop a program that will minimize wear on equipment so that breakdowns are at a minimum versus cost of maintenance.

In solving technical engineering problems, the engineer is trying to apply theory and principles. The effectiveness of the method applied is evaluated by the engineer to determine its degree of success, and if the method is unsuccessful, the engineer tries to determine why it did not perform as expected. In all of the engineer's work, he is establishing standards and then evaluating the results.

This characteristic of the engineer comes naturally and is always a part of his work. But as an engineer manager he must perform similar work with people; for example, he must make appraisal evaluations of his subordinates. This has many uncertainties and unknown factors, which many times cause the engineer manager to be reluctant to perform this very critical part of his job. Actually, the work is not so different, but the engineer manager is dealing with people rather than physical concepts and machines that are predictable whereas people are not. Fortunately, there are proven concepts for appraisal evaluation and personal development that are effective for engineers and that make these tasks enjoyable for the engineer manager if he uses them properly. To enable him to do this is the objective of this chapter.

One of the most difficult actions for an engineer manager is the appraisal of his subordinates. Because of this difficulty, he is reluctant to assume this responsibility and finds every excuse to avoid it. This widespread resistance to evaluating performance can be overcome, but first the manager must understand the real causes and problems in the appraising function.

PROBLEMS IN APPRAISING ENGINEERS

For the engineer manager, these have arisen from the standards employed in the appraisal system, the procedure for conducting the appraisal, and the fact that the factors being appraised were unrealistic. It is understandable that the engineer manager is reluctant to apply performance appraisals under such circumstances.

Effective evaluation requires that verifiable goals or objectives be established so that each employee knows to what his actual performance will be compared. Frequently, these goals or objectives are not established, or if they are, they are not communicated to the subordinates. It appears to the engineer manager that the appraisal function is not a well-developed science with proven techniques that can be assigned to an assistant. Further, it is a real problem for him to determine what should be measured or if it can be measured. Very often, factors are measured that cannot be defined in terms meaningful to the subordinate.

An impressive appraisal system can be established with all the necessary elements but still fail because it does not include the element of engineering performance. Many engineers have been evaluated against standards of personal traits and work characteristics rather than engineering activities. Such systems list personal characteristics as an ability to get along with people, analytical competence, initiative, and judgment. The work characteristics include ability to carry out assignments and instructions and knowledge of the job. Unfortunately, the people doing the actual appraisal tend to emphasize the personal traits instead of the work characteristics. The evaluation is not objective but rather a subjective judgment. As a result, top ratings are frequently given to everyone. The appraisal process soon becomes meaningless to both the subordinate and the superior. The engineer manager either resists the evaluation or tends to go through the motions, but many times he does not know how to rate factors that were developed and given to him by the personnel manager. This makes the engineer manager very uncomfortable because he feels that he is playing with people's lives, which he feels is not his responsibility as a professional engineer.

It is no wonder that the engineer manager thinks appraisals are something to be avoided and if someone else, such as the personnel department in the organization, would assist in appraising, he would prefer to let them do it. But good management dictates that appraisal is a responsibility of the manager and therefore he should learn how to do it.

THE APPRAISAL PROCESS   Performance appraisal is the systematic evaluation of the individual with respect to his performance on the job and his potential for development. During the past several decades there have emerged various approaches to performance appraisal.

The appraisal process is a managerial method concerned with the *what* and the *how* of a subordinate's performance. The appraisal also provides a means of satisfying the subordinate's needs for recognition and professional growth, and more important, a means of creating a strong relationship between the engineer manager and the subordinate, thereby causing greater job satisfaction and personal growth. The appraisal process is concerned with the needs of subordinates as well as the needs of the organization. Therefore, the appraisal process consists of two phases:

1. Continual review and appraisal of subordinates' results.
2. Periodic formal review and appraisal between engineer manager and subordinate.

Appraisal involves the day-by-day assessment of progress toward goals and objectives. The more frequently the engineer manager meets with his subordinates, the more day-to-day opportunities for appraisal. It gives him the opportunity to provide suggestions to his subordinates so that their overall performance will improve.

The effectiveness of the appraisal is a function of how many contacts the engineer manager has with his subordinate. It is useful for the engineer manager to review his contacts with the following checklist:

	Frequently	Infrequently
Discuss progress with subordinate	_____	_____
Discuss problems the subordinate is having with his work	_____	_____
Subordinate comes to discuss progress	_____	_____
Make comments to subordinate on his work	_____	_____
Provide information subordinate needs to do his work	_____	_____
Request information about subordinate's work activities	_____	_____
Encourage subordinate to discuss work activities	_____	_____

The more checks the engineer manager has in the "frequently" column, the more opportunities he has had for appraisal.

Over the years a variety of techniques have been developed for the appraisal of employees. Each of these will be discussed in some detail below.

APPROACHES TO APPRAISAL

1. *Comparison against standards.* The oldest form of these standards is based on personal traits and is still probably the most widely used of all appraisal methods. It may be referred to as a conventional rating skill technique. In practice, the direct supervisor of the employee being rated is supplied with a printed form. The form is normally grid-like and lists a number of trait characteristics with probably five ratings under each (Figure 8–1). The form for nonsupervisory employees is somewhat different from that for managerial employees. For example, in the case of nonsupervisory employees, the supervisor is more interested in the quantity and quality of work, job knowledge, cooperativeness, dependability, initiative, and attitude. In the case of managerial employees, he may be more interested in analytical ability, judgment, leadership, creative ability, knowledge of work, and emotional stability.

In Figure 8–1 the rater is given some explanation of the five grades. On other forms an average may be indicated in one column and varying degrees of above and below average may be indicated in four other columns. The rater may be asked to explain why he has chosen the particular degree of performance. This method forces the rater to think more carefully about his rating.

Although the rating skill technique using personal traits and behavioral characteristics is relatively easy to construct, it does have some disadvantages. Evaluation of personal traits is subjective in nature. Management is really con-

Job Knowledge

1	2	3	4	5
Serious gaps in his knowledge of essentials of job	Satisfactory knowledge of routine aspects of job	Adequately informed on most phases of job	Good knowledge of all phases of job	Excellent understanding of his job. Extremely well informed

Judgment

1	2	3	4	5
Decisions often wrong or ineffective	Judgement often sound but makes some errors	Good decisions resulting from sound analysis of factors	Sound, logical thinker	Consistently makes sound decisions, even on complex issues

Oral and Written Communication

1	2	3	4	5
Unable to express ideas clearly. Often misunderstood	Expresses ideas satisfactorily on routine topics	Generally expresses thoughts adequately	Consistently expresses thoughts clearly	Outstanding in written and oral expression

Attitude

1	2	3	4	5
Uncooperative, resents suggestions, no enthusisam	Often cooperates, often accepts suggestions	Satisfactory cooperation, accepts new ideas	Responsive, cooperates well, helpful to others	Excellent in cooperation, welcomes new ideas, very helpful & enthusiastic

Quantity of work

1	2	3	4	5
Falls below minimum requirements	Usually meets minimum requirements	Satisfactory quantity	Usually well exceeds minimum	Consistently produces superior quantity

Quality of Work

1	2	3	4	5
Poor quality, many errors or rejects	Quality usually o.k. Some errors or rejects	Satisfactory quality	Quality exceeds normal standards	Consistent high quality work

Overall Evaluation

1	2	3	4	5
Poor	Fair	Satisfactory	Good	Excellent

FIG. 8-1. A Conventional Rating Scale From Using Five Discrete Steps for Each Factor Being Rated. (Source: *Personnel,* Fourth Edition, by Dale S. Beach. Copyright © 1980 by Dale S. Beach. Reprinted by permission of Macmillan Publishing Co., Inc.)

cerned with job performance, and the degree of rating in various traits may not directly relate to this performance. It has been definitely proven that people who have different personalities may be equally effective in job performance. Further, the fact that there is a numerical rating gives the illusion of an accuracy that is not necessarily present. Since the engineer manager is accustomed to thinking in quantitative terms, he must be careful not to succumb to this illusion.

Another form of comparison against standards may be found in checklist methods. In these methods, a weighted checklist is used which contains a large number of statements that describe specific aspects of behavior that can occur in the job for which it has been designed. Each statement has a weight or score attached to it. The rater checks those statements which actually portray the behavior of the employee. Frequently the rater does not know the weights of the items and the scoring is done in the personnel office. Supervisors must think in terms of very specific aspects of behavior because separate lists are generally constructed for each job.

2. *Interpersonal comparisons.* Here the person being rated is ranked against others in his job or department on an overall basis. The supervisor is required to rank his subordinates according to their job performance and their value to the organization. Thus, someone is going to be at the bottom and someone is going to be at the top in the ratings. Normally, the rater will choose the best performer and the poorest performer first, and then group the remainder in between. He may also handle the in-betweens by picking first the best and then the poorest of the remaining, and so on. Another ranking system involves breaking a group into the lowest third, the middle third, and the highest third.

A more difficult approach involves comparing each employee with every other employee, one at a time. This is sometimes known as the *paired-comparison* technique. The top employee is judged by the number of times he was chosen over the other individuals in all the paired comparisons.

The major weakness of all these ranking methods is that they fail to reveal the differences between persons in adjacent ranks. Specific components of behavior are not analyzed; instead, the whole person is judged. In addition, there are no standards of performance against which these judgments are made.

An effort has been made to improve the rankings through the utilization of the normal frequency distribution curve. In using this method, the supervisor must allocate 10% of his people to the highest category, 20% to the next, 40% to the middle, 20% to below average, and 10% to the bottom. This process is questionable because in the small sample involved in the department there is little likelihood that the actual distribution of performance would match the normal distribution curve.

3. *Essay.* In this method the rater describes the performance, traits, and behavior of the employee in the form of a free-flowing essay. The rater may speak to the individual's strengths and weaknesses, leadership ability, technical effectiveness, and promotion potential.

Another variation of this method involves the development of a job description for the ratee. The supervisor then writes a paragraph or two stating how well the ratee has performed against his job description during the appraisal period.

Such an approach requires considerable time and thought. Since it has no quantitative scale, it is difficult to compare individuals. But it does have the advantage that in the review of ratings between the rater and ratee considerable data are available for discussion.

4. *Direct performance measures.* This appraisal method ignores the trait characteristics of the individual and directs itself to measurable output items. These may be defined as quantity, quality, attendance, sales dollars, etc. It is not as valuable in assaying the future potential of the individual. Also, it is difficult to assess the performance outputs as to what degree the results are the responsibility of the ratee and to what degree they were caused by circumstances beyond his control. For example, the salesman being rated might have had an occasion in which one of his major competitors had a plant destroyed by fire and was unable to meet the needs of his customers. As a result, the ratee might have enjoyed extremely high sales through no action of his own. For these reasons, such measurements of performance should probably be combined with some of the other rating systems.

5. *Management by objectives or appraisal by results.* In this case, each individual's performance is compared against his agreed-upon objectives. Management by objectives tends to overcome many of the problems of traditional appraisal systems. A major goal is to enhance the superior–subordinate relationship, strengthen the motivational climate, and improve performance.

A number of the key features of management by objectives are described below:

A. Superior and subordinate get together and agree upon the subordinate's principal duties and responsibilities. This is to develop a workable and understandable job description.
B. The subordinate and supervisor jointly develop the short-term performance goals that will be expected. The supervisor is especially concerned that the established goals will relate to those of the department and the entire organization. This is exceedingly important if all elements of the organization are to work together.
C. The supervisor and subordinate develop criteria for measuring performance against these goals. This requires considerable discussion and a decision must be made about whether or not quantitative measurements can be developed and used.
D. During the year the supervisor and subordinate watch the subordinate's progress toward the agreed-upon goals. They may informally discuss how this progress has taken place. Because they now have a common platform upon which discussion can take place, the relation-

ship between these two individuals is greatly strengthened. They may also find that circumstances beyond their control are going to affect the ability of the subordinate to meet his goals. In such a case, they can mutually agree to modify the goals to an attainable level.

E. Since the goals were developed mutually, the supervisor will be interested in helping his subordinate achieve his goals. It also gives the supervisor an opportunity to provide counseling if he sees particular weaknesses preventing the subordinate from achieving his goal.

F. In this appraisal process the supervisor is not so much a judge but instead finds himself in the role of assisting the subordinate in achieving his goals. This change in role will again help strengthen the relationship between the two and also improve the morale of the department.

G. This process tends to focus on the results that have been accomplished instead of on the individual's personal traits. Nevertheless, the supervisor will be concerned with those personal traits that may influence the subordinate's achieving his goals.

Douglas McGregor played an important part in the development of this procedure.[1] This method is recommended by many management consultants and is reaching an ever-stronger position in industry and educational institutions.

Developing Definable Goals

To measure performance, it is important to develop quantifiable goals if possible. These may be results that can be stated in terms of activity level, such as project completion, start of a project, or completing the installation of equipment. The results might be measured in very specific terms as illustrated in Table 8–1. To attain each of these goals requires a detailed plan of action. From these may be developed more specific goals for subordinates.

Management by objectives works exceedingly well for technical professionals and supervisory and executive personnel. Because of their background, education, and responsibilities, they work well with their supervisor in setting work goals. It is not a method that is applicable for hourly workers because their jobs are usually too restricted in scope.

This method of appraisal has a number of advantages. By participating in setting goals, the individual has a much stronger interest in meeting them. In a sense, the supervisor and the subordinate are on the same team. The subordinate's goals are set and evaluated in terms specifically suited to his situation and abilities. Defensive feelings are minimized, because the superior is not acting as a judge but is on the same team. Since the emphasis is on performance rather than personality traits, the subordinate is less defensive. This form of appraisal is also more exciting to both parties because its emphasis is on the present and the future, areas over which they have some control. Other appraisal methods reviewed tend to point to the past.

[1]Douglas McGregor, "An Uneasy Look at Performance Appraisal," *Harvard Business Review* (May–June 1957), 89–94.

TABLE 8–1

Definable Goals Against Which to Measure Results

Cost: Reduce direct labor by 5%
Profit: Increase return on investment by 2%
Marketing: Increase product X's sales by $500,000 per year
Product development: Redesign product Y line during next 12 months
Personnel development: Reduce turnover by one-quarter
Engineering management development: Have each engineer enroll in
  one short course related to future projects
Product liability: Install X-ray equipment for detecting product failures

But like all appraisal systems management by objectives has its disadvantages. Some managers find it difficult to work in such a participative atmosphere. Also, the subordinate may press toward setting easily obtainable goals so that he will look good when compared against them. And there is a tendency to quantify the goals, perhaps stressing too much such items as profits, cost, and efficiency. Investments in human resources may get downgraded.

Also, it is not as helpful in deciding salary increases and promotions. It is difficult to compare one individual with another, since the basic thrust of the procedure is toward comparison against standards. These standards are personal and are not uniform throughout the department. Since the objectives tend to be short range, they do not identify those characteristics and traits that might qualify one for a future promotion. Carroll and Tosi have done an excellent job of summarizing the conclusions of many investigators of the MBO process.[2] They found that the degree to which performance improved depends whether or not the goals are at the proper level of difficulty for the individual, whether or not proper time limits are set, and whether or not the goals are specific.

To assist in the execution of the MBO-type of appraisal, a form such as illustrated in Figure 8–2 can be useful. During the initial meeting between the supervisor and his subordinate the left-hand column can be filled out with the assigned activities. The second column, planned actions, can also be developed at that time. When they meet again, perhaps a year later, the results can be tabulated in the third and fourth columns. These columns can be used to explain deviations and to indicate corrective actions to be taken during the following period. At that particular time a sheet can be prepared listing the assigned and planned actions for the ensuing year. In this manner, continuing data are available on the individual's progress toward meeting his objectives.

The engineer manager must approach the meeting with his subordinate with an open mind. He must look upon it as an opportunity to strengthen his relationship with the employee, to evaluate the employee's ability to perform, and to evaluate his potential for advancement.

It is, therefore, important that he reassess the goals that were established

[2]Stephen J. Carroll, Jr., and Henry L. Tosi, Jr., *Management by Objectives: Applications and Research* (New York: Macmillan, 1973), Chap. 1.

Name _____

Position _____

Appraisal period from _____ to _____

Assigned Activities	Planned Actions	Results	Recommended Changes
		Results are measured against planned activities	Note reasons for deviation, corrective action taken, and modification of objective if necessary

FIG. 8-2.   Appraisal Worksheet

the year before and look carefully at whether or not intervening factors have had any influence on the subordinate's ability to meet these goals. If so, he should take the initiative of pointing this out during the course of the interview. He also should review whether or not the actions taken by the engineer during the past year have contributed to the organization's objectives. This gives him an opportunity to communicate with his subordinate on how he has contributed to these larger goals. If the goals are met or exceeded, this is an opportunity for the manager to compliment the employee on his contributions to the overall corporate success. He can encourage him to continue on in this vein and perhaps outline for him something of his future with the organization.

If the goals have not been met, the manager must first determine why and decide if it is a personal failure on the part of the subordinate or if it was impossible to have achieved the goals. If there are personal corrective actions needed, he must communicate these in such a manner that the employee will want to improve during the next period so that he can be looked upon favorably by the manager. This is an opportunity to strengthen the respect that his subordinate holds for him through a frank discussion of any weaknesses that have developed during the preceding year.

As the interview is concluded and the goals for the following year are set, the manager should take advantage of the opportunity to outline the growth expected in his department, in the company, and the challenges that are ahead. Since the engineer is motivated by challenges, placed in the proper light these can be significant motivators.

During informal meetings throughout the year the manager should be constantly encouraging the subordinate to practice self-appraisal in terms of meeting goals. He should get across to the employee that by doing this he is acting like one of the control valves in the plant, in that he is providing constant feedback to correct his progress toward goals. The manager can encourage this by inquiring from time to time about the progress that is being made, not from a critical but from an information standpoint.

Finally, the engineer manager must determine the rewards that he is going to provide for outstanding performance. These may fall in the area of salary increases or recommendations for promotion. Or he may find that the individual has excellent potential but needs some additional training in order to perform at higher job levels. This underscores the last phase of the definition of appraisal, which was to evaluate "his potential for development."

TRAINING AND
DEVELOPMENT   As previously mentioned, the appraisal function is a very important part of the engineer manager's responsibilities and deserves a considerable amount of his time and effort. However, after the appraisal evaluation is performed, the engineer manager must follow it up with an effective training and development program to correct the areas in which performance was unsatisfactory. Today, more than ever before, the manager must assume the responsibility of training and developing his people so that each one contributes his full measure to the organizational objectives and becomes a productive employee. Somehow, the generally accepted viewpoint is that one can acquire the necessary skills or techniques to improve himself without being given even the basic instructions. Some feel that, given sufficient time and experience, the individual will eventually correct or overcome his weaknesses. But this is really wishful thinking. Training and development must be an integral part of every job, not a separate activity that can be accomplished when time is available.

Objective of Training and Development

Engineers and other technically oriented people are concerned about their work from two viewpoints: "How am I doing?" and "How do I improve?" An effective appraisal system will answer the first question and an effective training and development program will answer the second question. The major objectives of the organization's training and development program are to provide the engineers and other employees with an opportunity and the motivation to achieve their maximum productivity. The direct benefits of the program will provide the individual with the feeling of achievement, will give him satisfaction, and will thereby increase his productivity, earning ability, and security. In addition, the organization will benefit from increased productivity.

Another basic reason for training the organization's engineers is the realization that any four-year engineering program cannot provide an individual with the breadth and depth required for most engineering positions in the organization. The engineer's formal training and background do not usually match the diverse needs of the particular department. Also, all areas of engineering are rapidly changing, which requires engineers to be updated on new concepts and techniques or they may become obsolete.

Evaluating the Needs for Training and Development

Training and development are progressive processes in the sense that they should not be thought of as something that can be completed; there are no limits to the degree a subordinate may be trained and developed. Training involves learning skills to improve the efficiency in the achievement of a particular objective. In evaluating the needs, it is essential for engineer managers to first determine what must be done. This is based on the organization's goals or objectives. Before any "needs" evaluation can take place, one must know the firm's goals or objectives and what must be done to accomplish them.

Second, each individual must be realistically evaluated in terms of performance and needs. In such evaluation, there is a tendency for the engineer manager to concentrate on overcoming individual "weaknesses," which is a *negative* approach. A *positive* approach should be taken, one that emphasizes training and developing the individual's "strengths" and then correcting his weaknesses.

The third step in the evaluation process is to determine what management wants its people to become and what new skills and abilities must be acquired to achieve goals and objectives.

The fourth step involves comparing the people available and their skills with what people "need" to be.

Fifth is the development of a list of training requirements and developments that have been initiated in the process.

Sixth is the prioritizing of the requirements and scheduling, so that one activity builds on another and the most critical activities are undertaken first.

Seventh is the determination of the engineers' attitudes toward the proposed training and development program. It also includes providing motivation for subordinates to learn skills, new behavior, and techniques.

The eighth step is to perform follow-up evaluation on a regular basis to determine each activity's effectiveness in adequately meeting the needs of the group.

Determining the training needs based on an evaluation of performance necessary to achieve stated goals and objectives provides the engineer manager with a training program in areas in which there is a potential for improvement. To perform the evaluating steps, past performance records should be used because they give many clues to individual skills that need development. Also, tests, direct questionings, and observation all identify training needs. Thereby, a training program for improvement can be developed that is tailored to the unique requirements of the organization. Then the training program can be directed toward providing employees with the means to develop the necessary skills and abilities.

The ninth step is rewarding the subordinate for accepting training and development. Because the goals and objectives are constantly changing and because people are constantly changing, evaluating the "needs" for training and development must become a regular practice and function for the engineer manager. If not, his training and development program will soon be obsolete and his people will become reluctant and unmotivated to participate and unable to help achieve the organization's needs.

TRAINING PROGRAMS The engineer manager has both formal and informal methods for his engineers, technicians, and other employees to acquire and apply knowledge in a practical, effective manner. He must provide opportunities for all of his people to develop to their maximum potential so that they are motivated to excel. He must assist his people in many different ways in order to take advantage of available training opportunities. Following are descriptions of the most frequently used methods:

1. *On-the-Job-Training.* This technique is used more for training than development, but it is very popular with many engineer managers. The objective of on-the-job-training is to expand the individual's knowledge about the operations of the organization and the basic problems facing these various operations. On-the-job-training may take many forms such as:

A. Individuals may be assigned to predetermined jobs in various departments for a given work period. At the end of the period, each has had experience with a diversified number of activities. The objective may be to provide the individual with experience in as many activities of the firm as possible. In actual practice, serious problems develop in this type of on-the-job-training. Many times the assignments are of short duration. Since the trainee is primarily observing, his contributions are limited, and thus the program is expensive for the participating departments. There is also a question of the value to the individual's personal development if the work assignment in a given department is only for a few weeks or months. Two other problems frequently arise: (a) When positions open in the middle of the program, should the individual leave the program or continue to the end? (b) If the individual completes the program and no positions are open, what should be the next phase of his personal development program?

Aside from these objectives, this type of on-the-job-training does allow the individual to observe the department's activities. He has the opportunity to ask questions about the department's problems and operations. And it gives the manager an opportunity to explain his department to prospective subordinates.

B. Another type of on-the-job training is that in which certain *development positions* are included in the organization and are filled on a regularly scheduled basis by the individuals being trained. These are a regular part of the organization and provide the individual with actual experience. These positions are fewer in number so that the trainee receives a greater depth of experience and less breadth.

There is a problem in identifying appropriate positions that can operate with constant turnover, ones in which the new person learns his duties quickly, subordinates are experienced, and activities are relatively standardized. In any case, employees of these departments often resent having a temporary individual, and they feel that qualified people in the department are passed over for these positions.

To overcome some of the above problems, a common practice is to move promising individuals from one department or activity to another, which can be horizontal moves in the organization. These are not specified as on-the-job training but are set forth in a formal personal development program for the individual. The individual is not told how long he will be in the position; it may be for a short period or be rather permanent if he does not show signs of growth and development. There is no commitment that a change will take place at some future date. This practice avoids the problem of reserving special positions for training purposes or for holding positions open for individuals until they complete a program. It also prevents the trainee from feeling that he is only temporarily in a position.

2. *"Assistant to" Position.* Often the "assistant to" position is used in a personal development program. It provides on-the-job training experiences and exposure to many different problems and activities. In practice, these can be valuable experiences for developing an individual for future positions. The success of the position depends heavily on how effectively the manager provides training and adequate counseling time to this function. He can make the position meaningful by giving assignments that have value. For example, the assistant can be appointed as acting manager when the engineer manager is on vacation or on out-of-town trips. These assignments give the engineer manager an opportunity to evaluate the individual's actions and they provide an excellent development opportunity for the individual.

3. *Formal Training.* One requirement of a successful training program is the correct selection of a training method that can develop subordinates' skills and abilities. There is a wide variety of formal training methods available. Table 8–2 indicates the effectiveness of various training methods in meeting training objectives. It will be noted that programmed instruction is the most effective method of

TABLE 8-2

Effectiveness of Training Methods

Training Method	Knowledge Acquisition		Changing Attitudes		Problem-Solving Skills		Interpersonal Skills		Participant Acceptance		Knowledge Retention	
	Mean	Mean rank	Mean	Mean rank	Mean	Mean rank	Mean	Mean rank	Mean	Mean rank	Mean	Mean rank
Case study	3.56^b	2	3.43^d	4	3.69^b	1	3.02^d	4	3.80^d	2	3.48^e	2
Conference (discussion) method	3.33^d	3	3.54^d	3	3.26^e	4	3.21^d	3	4.16^a	1	3.32^f	5
Lecture (with questions)	2.53	9	2.20	8	2.00	9	1.90	8	2.74	8	2.49	8
Business games	3.00	6	2.73^f	5	3.58^b	2	2.50^e	5	3.78^d	3	3.26^f	6
Movie films	3.16^g	4	2.50^f	6	2.24^g	7	2.19^g	6	3.44^g	5	2.67^h	7
Programmed instruction	4.03^a	1	2.22^h	7	2.56^f	6	2.11^g	7	3.28^g	7	3.74^a	1
Role playing	2.93	7	3.56^d	2	3.27^e	3	3.68^b	2	3.56^e	4	3.37^f	4
Sensitivity training (T-group)	2.77	8	3.96^a	1	2.98^e	5	3.95^b	1	3.33^g	6	3.44^f	3
Television lecture	3.10^g	5	1.99	9	2.01	8	1.81	9	2.74	9	2.47	9

[a]More effective than methods ranked 2 to 9 for this objective at 0.01 level of significance.
[b]More effective than methods ranked 3 to 9 for this objective at 0.01 level of significance.
[c]More effective than methods ranked 4 to 9 for this objective at 0.01 level of significance.
[d]More effective than methods ranked 5 to 9 for this objective at 0.01 level of significance.
[e]More effective than methods ranked 6 to 9 for this objective at 0.01 level of significance.
[f]More effective than methods ranked 7 to 9 for this objective at 0.01 level of significance.
[g]More effective than methods ranked 8 to 9 for this objective at 0.01 level of significance.
[h]More effective than method ranked 9 for this objective at 0.01 level of significance.

Source: Stephen J. Carroll, Jr., Frank T. Paine, and John J. Ivancevich, "The Effectiveness of Training Methods," *Personnel Psychology* (Autumn 1972), 498.

acquiring and retaining knowledge. For changing attitudes and developing interpersonal skills, the sensitivity laboratory training method is most effective. Conference methods are effective in gaining participation, and the case method is best for acquiring problem-solving skills. However, the success of most training programs is directly dependent on the individual who is instructing and on his ability to sense the participants' needs.

Motivating Employee Training

The engineer manager must recognize that any training program involves changing the participant's behavior and attitudes and that the trainee must be interested in this change. One major responsibility of a manager is to create a desire and interest for his people to change their behavior. He must demonstrate that the training program will be in their own best interest, resulting in promotion, greater job satisfaction, and higher salary. This means he must follow up on the progress of the trainee who has completed the program and show his personal interest.

The concepts of training are fairly well developed to assist the engineer manager in developing an effective training program to meet his objectives. These can be stated as follows:

1. People are more responsive to training programs when they understand the purpose of training in terms of:
   A. Problems it will correct
   B. Benefits
   C. Relation to present performance
2. Rewards should quickly follow when desired performance is achieved.
3. Rewards for good performance following training must be sufficient.
4. Punishments have a disruptive influence on training.
5. Training should be followed with evaluation and corrective action should be taken.

Training is generally most successful when rewards are used to support a training program. Any activity will be effective for individuals when it gives pleasure to them, and it will be avoided when little benefit is derived from it. Therefore, if satisfactory rewards are received either during or after training, additional training will be accepted and an effort will be made to use the new techniques that have been learned; but if no feeling of satisfaction is received, future training will be avoided. During training, the more practice that is involved, the more likely the new technique will be learned.

All training programs should have some form of feedback during and after the training program. Frequently, tests are given before and after the training program and again six months later. The objective is to determine whether changes in knowledge, attitude, and abilities have in fact taken place. Feedback from these analyses of the trainee provides reinforcement to the training and is valuable as a motivator.

PERSONAL DEVELOPMENT PROGRAMS FOR ENGINEERS

The engineer is generally very receptive to personal development, but the program, to be effective, must benefit the engineer as well as his company. The following basic concepts should underlie the personal development program for the engineer:

1. The engineer must recognize the need for development. This requires the manager to provide information and counseling so that the engineer will recognize his individual needs.

2. He must be motivated to be trained. The extent to which he is motivated depends on the environment that the manager has created. This requires the manager to work with each individual, helping him to see the value of training.

3. His personal development program must be planned by the engineer manager in cooperation with the individual engineer. This involves short- and long-range planning so that each engineer's personal development needs complement the organization's needs.

4. Emphasis must be on immediate development that improves the engineer's present performance and on helping to achieve future development plans.

5. All individuals in the organizational unit must be given the opportunity to enter a personal development program.

6. The engineer manager must provide leadership in planning for the personal development of his subordinates.

7. Personal development requires a combination of formal training and on-the-job experience.

8. The personal development program must be tied to performance appraisal.

ORGANIZATIONAL DEVELOPMENT

In recent years organizational development has received considerable attention as an approach to achieve significant long-term changes in the organizational personnel.[3] Organizational development attempts to change attitudes toward a greater degree of participative management and away from authoritarian management. Once this change has taken place, restructuring of the organization can occur.

Organizational development requires the engineer manager to determine what type of group he wants to have. If he follows an organizational development program completely, a new organization will be developed, one characterized by a high degree of participative and decentralized management in which subordinates define their own work structure and plan their own activities. The real question for the engineer manager is whether or not this is the type of organization he wants to develop over a period of time.

Effective Appraisal of Performance and Personal Development Training

The conditions for effective appraisal of performance and personal development training might be summarized as follows:

1. The engineer manager must have definite organizational goals and objectives before appraisal and development can properly take place.

2. All levels of management must be committed to an appraisal evaluation program and development training.

3. The manager must be committed to a continual program of appraisals and development (not just when convenient).

---

[3]Warren G. Bennis, *Organization Development: Its Nature, Origins, and Prospects* (Reading, Mass.: Addison-Wesley, 1969), pp. 2–3.

4. The personal development program can be successful only when based on an effective appraisal evaluation system, which in turn must be based on having developed meaningful goals or objectives.

5. Training programs must be planned so that each one builds upon previous ones in a logical, consistent pattern.

6. Training must relate to present positions.

7. Training opportunities must be made available to all.

8. Training methods need to be tailored to the subject and material being presented.

9. Because of differing abilities, backgrounds, experience, and motivation, each individual will learn at a different rate. Therefore, personal development must be adapted to the individual and not to a group.

10. Learning requires seeing, hearing, and doing, making it necessary to use a variety of training techniques.

11. The engineer manager has the responsibility of motivating and guiding his subordinates to undertake personal development.

12. Most personal development training requires a combination of on-the-job and formal instruction.

13. Good performance must be rewarded immediately.

One of the most important functions of the engineer manager is **SUMMARY** to appraise his subordinates' performance and potential for development.

A number of techniques have been developed over the years, including comparison against standards, interpersonal comparisons, essay, direct performance measures, and management by objectives. In the application of these techniques, a number of procedures are used. One of long standing involves the utilization of forms which measure the various levels of behavior of the employee on numerical scales. These are easy to use, but they have the disadvantage that the judgments are subjective and they tend to look at the individual rather than at his performance.

Another technique involves rating each individual against other individuals in the department and providing a ranking. When this is done, one individual ends up at the top and another at the bottom. Other approaches to ranking compare each individual against every other individual in the department. The major weakness of this method is that it does not reveal the amount of difference between persons in adjacent ranks and it does not consider specific components of behavior. Standards of performance are not involved.

The essay has also been used as a method of reviewing performance. It can deal with the strengths and weaknesses of the individual, or it may deal with his performance against a previously developed job description.

Performance can also be measured against direct outputs such as quantity of production, quality of production, attendance, or sales dollars. The disadvantage of this technique is that it tends to ignore the individual. Also, his performance may be influenced by factors beyond his control.

The management by objective system is probably gaining most in acceptability today. It has the added advantage that it strengthens the relationship between the subordinate and his supervisor, for the supervisor does not sit in judgment. Together they agree on goals and together they review the progress toward the goals. Its disadvantage lies in that it is not as useful for salary decisions because it doesn't compare individuals with one another. It also tends to ignore the personal characteristics of the individual.

The appraisal interview offers an excellent opportunity for the supervisor to strengthen his relationship with his subordinate and to counsel him toward future improvements. It also allows them to mutually work out the goals for the succeeding period based on what has happened during the prior year. The supervisor has the opportunity to boost the employee's morale by helping him to see where the organization is going in the future.

One of the most important objectives of the appraisal system is to sort out those individuals who need training and development. They are then placed either in on-the-job training programs or special formal training programs. The goal of these programs is to prepare them for doing a better current job or for future promotion.

Important Terms

Interpersonal Comparisons:	Ranking of various subordinates against each other either within a department or on a broader corporate basis.
Management by Objectives (MBO):	A system of performance appraisal involving the determination of goals jointly by superior and subordinate and subsequent measure of performance against these goals.
Performance Appraisal:	A systematic evaluation of the individual's performance on the job and potential for development.
Standards:	Descriptions of various levels of behavior against which subordinates can be compared.
Training and Development:	Through on-the-job training or formal training, improve the employee's current performance or prepare him for future promotion.

For Discussion

1. Why is it so much harder to appraise people than to appraise machines and equipment?
2. "The evaluation of an individual is not objective but is only a subjective judgment." Comment and discuss the implications of this statement.
3. Why is it so hard for the engineer manager to appraise his employees?
4. Who should do the appraisal? Why?
5. List five approaches to appraisal.
6. Describe comparison against standards.
7. How often should appraisal evaluations be made? Why?
8. Describe methods that can be used to evaluate each employee relative to another. What are the advantages and disadvantages of this technique?
9. What steps are involved in determining the training needs of an organization?
10. What are the advantages and disadvantages of direct performance measures?
11. Why is personal development important, and what should be involved in the personal development program?
12. What is a major problem with the on-the-job training technique?
13. "It is critical for the engineer manager to make effective appraisals." Make as detailed comments as possible on this statement.
14. Training and development are common practices today in engineering management. List the important factors in evaluating the need for training and development. In what ways can it be effectively carried out?
15. What does setting goals have to do with the appraisal process?
16. Describe the steps involved in an effective management by objective appraisal.

Case 8–1

The Performance Appraisal

Randy, an engineer in R&D made a very serious mistake. In the design specification he prepared for the Midwest Equipment Company he made a mistake in a basic calculation. The equipment was ordered against these specifications, has been shipped, and is being installed by a local contractor. The contractor questioned the size of the equipment and the salesman questioned the specifications on the equipment he was installing.

It is the policy of R&D to have someone else check the work and approve it before it is sent over to sales who gives it to the customer. Normally, the checking engineer quickly reviews the calculations and initials the specifications and sends them on to sales. Each has more to do than he can handle and all are behind schedule; therefore, they cannot devote much time to this operation.

When sales checked into the contractor's question the error was discovered. Rex, of course, was called in about the error. Talking to Randy, he said, "You really made a serious mistake on the Midwest Equipment Company design, and it is going to cost this company plenty!" Then Rex explained what had happened.

A few days later Rex received a note from the personnel department stating that Randy's performance appraisal was due the end of the month. Rex thought this would be a good time to stress Randy's poor performance.

On Wednesday Rex called Randy into his office. "Have a seat. It's time for your annual performance appraisal. As you know, your work has gotten worse during the year and it seems each time you make bigger mistakes."

"Oh, I thought I was doing better," said Randy.

"Better! After last week's foul-up with the Midwest Equipment job, how could you say that?" asked Rex.

"Yes, but that was just a bad day. Remember on the H & H job you even received a letter stating how much money we saved them through the specifications I wrote. After that you even told me I was doing better," said Randy.

"I said your attitude was better and that you were working better with others, not your work," said Rex.

"What else have I done wrong?" asked Randy.

"You should be showing more initiative and imagination in your work," replied Rex.

"I work harder and spend more time after hours than anyone else. I always stay when you ask me, which is more than the others do. Remember the Saturdays I have come in here to work?" reminded Randy.

"But last week's mistake was inexcusable," said Rex.

"Look, Rex, I told you two months ago that we did not have time to check each other's work properly and that one day something like this would happen. I told you so!" declared Randy.

"The point is your work must improve. I must fill out your performance appraisal and it is not going to be good!" Rex remarked.

"That's not fair, and it's not ethical. My work is better than the other engineers," said Randy.

"I am rating *you,* not the other engineers," was Rex's response.

1. How objective will Rex be in this appraisal?
2. What do you think of Randy's conduct in the interview?
3. Will Randy's performance improve?
4. How should Rex have handled the appraisal and what should he do in future interviews with Randy to get better results?

Case 8–2

Training and Development

About two years ago, at the urging of Rex Smith, Tom Johnson, and others, the company management started an extensive management development program. Top management was sold on the fact that the program would benefit the company by providing a group of qualified individuals to fill vacancies and therefore not require the company to go outside to fill vacant positions.

The personnel manager encouraged people from all levels of management to participate in the program to improve their management skills and to qualify them for future promotions within the company.

The program involved company in-house short courses, short courses at universities, and evening college credit courses. The program for an individual was developed in cooperation with the personnel educational staff and generally included all three types of courses. These courses were presented at company expense.

Tim Casey, one of Rex Smith's outstanding engineers, worked out a program of study with the personnel department. When the program was initiated, he worked hard and did extremely well. Also, he encouraged others to participate in the program. Tim felt that his efforts would be rewarded through promotion when an opening occurred.

Tim applied twice for positions in manufacturing, but in each case outsiders were hired to fill these positions. Tim felt that his experience plus the additional knowledge from the management development program made him better qualified than the individuals hired. He asked Rex to find out why he was not promoted. When Rex inquired, he was told that in each case no one in the company was considered qualified for these openings. Also, Rex was told that it takes a long time to develop an individual for higher level responsibilities in manufacturing. Rex has also heard that other program participants who applied for vacant positions had been turned down and an outsider hired. As a result, most participants now have the attitude that the program is a waste of time because the knowledge and experience gained are not recognized by the company management.

Rex must explain to Tim why he was not promoted.

1. How should Rex explain the situation to Tim?
2. Do Tim and the others have a valid complaint?
3. What recommendations, if any, should Rex make about the operation of the training program to the personnel director?
4. What moves should top management make to correct the situation?

For Further Reading

BEACH, DALE S., *Personnel,* 4th ed. New York: Macmillan Publishing Co., Inc., 1980.

DUNN, J. D., and FRANK M. RACHEL, *Wage and Salary Administration.* New York: McGraw-Hill Book Company, 1970.

GUSTAT, G. H., "Incentives for Indirect Labor," *Proceedings,* Fifth Industrial Engineering Institute. Los Angeles: University of California–Berkeley, 1953.

KELLOGG, MARION S., *What to Do About Performance Appraisal,* rev. ed. New York: American Management Association, 1975.

KOONTZ, HAROLD, *Appraising Managers as Managers.* New York: McGraw-Hill Book Company, 1971.

———, CYRIL O'DONNELL, and HEINZ WEIHRICH, *Management.* New York: McGraw-Hill Book Company, 1980.

LANHAM, E., *Job Evaluation.* New York: McGraw-Hill Book Company, 1955.

LOVEJOY, L. C., *Wage and Salary Administration.* New York: The Ronald Press Company, 1959.

ODIORNE, GEORGE S., "MBO in the 1980's: Will It Survive?" *Management Review* (July 1977), 39–42.

"Optimum Use of Engineering Talent," *AMA Management Report #58.* New York: American Management Association, 1961.

PATTON, ARCH, "How to Appraise Executive Performance," *Harvard Business Review* (January–February 1960), 63–70.

RIEGEL, JOHN W., *Administration of Salaries and Intangible Rewards for Engineers and Scientists.* Ann Arbor, Mich.: The University of Michigan, 1958.

ZOLLITSCH, HERBERT G., and ADOLPH LANGSNER, *Wage and Salary Administration,* 2nd ed. Cincinnati: South-Western Publishing Company, 1970.

PARTICIPATIVE MANAGEMENT TECHNIQUES FOR THE ENGINEER MANAGER

One of the first prerequisites for gaining acceptance of the engineer manager's ideas is obtaining the participation of the engineers. To obtain the support of the engineers requires the engineer manager to develop effective participative management techniques.

CHAPTER HEADINGS

- ☐ Types of Participative Management
- ☐ Developing a Participative Strategy
- ☐ Achieving Participation Through Involvement
- ☐ Some Common Errors in Participative Management
- ☐ Implementing Participative Management
- ☐ Constraints and Barriers to Participative Management
- ☐ Summary

LEARNING
OBJECTIVES

☐ Gain insight into the role of participative management in the organization.

☐ Learn about the informal and semiformal methods involving participation.

☐ Understand the methods of achieving participation through involvement.

☐ Learn methods for implementing participative management.

☐ Become acquainted with some of the limitations and barriers to effective participation.

EXECUTIVE COMMENT

J. HAROLD YEAGER
Senior Vice President
Corporate Technical Services
Mallinckrodt, Inc.

J. Harold Yeager joined Mallinckrodt in 1942. He served in the former Uranium Division until 1964 when he was appointed technical director of the Operations Division. He was elected a corporate vice president and named general manager of that division in 1967. He assumed responsibilities in the Chemical and Specialties Group and in 1978 was promoted to senior vice president, with his current responsibilities.

The Importance of Participative Management

An engineer manager should possess skills in participative management techniques to achieve his full potential as a manager. The applications for these skills will vary, depending on his job assignment.

One assignment could be to manage other engineers, technicians, or draftsmen in performing technical tasks. Examples include managing a drafting group, an instrument design group, or a mechanical design group. Usually, the task is well defined, and the manager's job is to direct and guide his people toward the achievement of an acceptable design. The skills of participative management are useful to achieve creativity and design control within the limits of the assignment. In these examples the engineer manager's use of the technique does not differ from that of the usual manager of another function.

Another assignment could be as a project manager. Such a person is usually assigned the task of providing a facility for achieving some new or different operating result. Projects can range from buildings for warehouses and offices to complex manufacturing processes. The project manager must obtain data, decisions, and commitments from many functions inside and outside the organization. For example:

1. The marketing/sales function must commit to specific opportunity objectives.
2. Research usually identifies the technology to be used.
3. Production must commit to cost and capacity objectives.
4. Financial usually consolidates or reviews financial data to keep the project in line with company objectives.
5. Service organizations in a plant, such as utilities, maintenance, and quality control, must be considered.
6. Staff departments, such as safety, personnel, and insurance, will usually have suggestions or requirements affecting design or operating modes.
7. Specific designs can be accomplished by an inside group or by an outside firm.
8. A general contractor or a group of contractors usually builds the facility.
9. Outside groups can influence design and ability to operate; for example, Environmental Protection Agency, Occupational Safety & Health Administration, Toxic Substances Control Act, Department of Energy, Nuclear Regulatory Commission are familiar agencies whose regulations must be considered.

Participation in all of these functions, both inside and outside the organization, is essential to the successful execution of a project. The project manager's job requires a high degree of coordination, usually with the attendant requirement of resolving conflict. Participative management techniques are the tools which the skillful engineer uses to keep his project on course and to assure good performance. The executive who accepts ultimate responsibility for the success or failure of the project usually looks to the project manager as the key person who will either resolve the conflict or bring such issues to him for resolution.

The project manager who obtains good performance will have all participants sharing in the good results. Projects which fail usually wind up on the shoulders of the project manager; thus it behooves him to have successful participants in the project so that he may join in their success rather than bear the loneliness of failure.

The third assignment could be as the manager of a corporate or divisional engineering staff where management functions such as design, project management, and construction management are combined. This manager must first see that a participative climate exists.

Second, he must be sure that project managers are asking for and receiving needed input from all of the functions in the organization which are involved in any project. Third, issues will arise which can only be resolved between the head of corporate engineering and the appropriate divisional head. The development of skills of participative management within the engineering department will reduce the number of conflicts which he must resolve, and his own personal skills will help to resolve the remainder.

One final thought: Participative management is a term that covers well-defined managerial procedures and techniques. It should not be confused with committees or other group activities. The engineer must be held accountable for achieving the result needed by the operating group. Skills in the use of participative management techniques can be among his most valuable tools.

Whatever his duties or responsibilities, the typical engineer of today is surrounded by a complex technological environment. This requires him to reconcile systems and procedures, understand and execute company policies, and develop essential technical skills to perform his duties properly. This is a difficult task which becomes complicated with many roadblocks. In many cases, he finds himself reporting to several superiors. The result has been that for some engineers a job is "only a job." He must work to earn a living and has resigned himself to this type of environment. As a result, the engineer loses interest in both his work and his company. Some engineers are always looking for a job, or just making minimum contributions to the organizational goals, or constantly complaining about everything. Some engineers, however, are very enthusiastic about their work, always willing to take on responsibilities. There are infinite numbers of answers to the question of why this is so. We will examine what may be done with the environment in which the engineer works that will cause him to be a willing and enthusiastic team member. One method of doing this is known as participative management.

Participative management may be defined as the process by which people contribute ideas toward the solution of problems affecting the organization and their jobs. Using participative management changes the conventional relationship between the engineer manager and his subordinates in various ways. The predominant one is the increase in the amount of influence his subordinates have in the direction of his department (Figure 9–1). Participative management techniques involve the individual in the problem-solving process. Participation may range from full involvement to merely offering suggestions.

The degree of participation at any given time is determined by (1) the type of problem and (2) the willingness of people to get involved and actually participate in the problem-solving process.

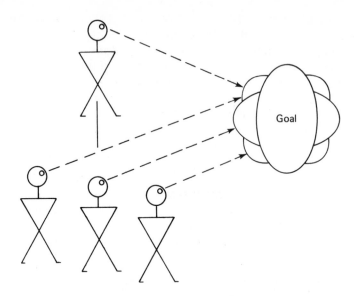

FIG. 9-1. Participation Occurs When Both Manager and Subordinate Understand, Accept, and Work Toward the Same Goal.

The basic reasoning underlying participative management is that all individuals have a need for recognition. People want to be involved, considered, and feel that they are worth something as indicated by the feedback relationship in Figure 9-2. The objective of participative management is to develop feelings among employees that lead to a common commitment to the goals and objectives of the group. Participation provides an open discussion of issues and feelings and promotes constructive cooperation.

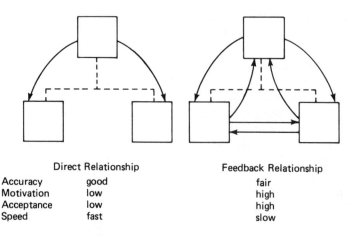

	Direct Relationship	Feedback Relationship
Accuracy	good	fair
Motivation	low	high
Acceptance	low	high
Speed	fast	slow

FIG. 9-2. Communication Relationship

When managers establish means on either an informal or a formal basis for obtaining help from subordinates in making plans and decisions, they are tapping the knowledge and creativity of others. In this manner, managers can obtain valuable advice and assistance from their subordinates, since they cannot possibly know all the problems and issues connected with work in their department. This participation brings into play the higher drives and motives of the employee. The opportunity to satisfy drives for self-expression and accomplishment, autonomy, and self-assertion lets the employees know that their contributions are sought and appreciated.

Participative management is often considered the opposite of authoritarian management. However, low participation does not mean that an authoritarian management style occurs; it simply implies individual decision making by formal authority of the position or by policies and procedures developed through formal authority.

Unfortunately, there is much misunderstanding and misuse of the participative management technique. Consequently, many times participative management ends up to be more fiction than fact. Some engineer managers think that they are practicing participative management when they give pep talks to their engineers about working together or try to persuade their people to work more enthusiastically. All the engineer manager is trying to do in these cases is inspire people to do their jobs; this is not the participative management process of problem solving. The strategy of participative management is involvement.

It must be pointed out that the autocratic style is not necessarily bad as opposed to the participative management style. Certain situations require the engineer manager to provide the autocratic style of leadership. As mentioned earlier under leadership, the manager needs to alter his style to best fit circumstances at a given time.

Participation is appropriate for all levels in the organization hierarchy. In practice, it occurs from first-line supervisory levels to the top of the organization.

It is worthwhile to consider the differences between participative management in business and democracy in government. In a democracy the citizens set up their own governing body and make their own laws through elected representatives. People have the power to elect or reject their representatives and leaders. In the work organization, however, the employees do not select their leaders, since they are appointed by management at the top of the organization. If the manager is ineffective as a leader, he is removed by his superiors in the organization and not by the employees. Therefore, when the manager consults with his subordinates and shares some of his decision-making authority with them, he does this voluntarily. He still retains the final authority and most of the power to make and implement these decisions. He can also rescind his sharing of the decision-making power at any time.

In practice in industry today, the use of participative management is discretionary on the part of management. Management has the power to create such a program and to abandon it. The major exception to this would be in the area of collective bargaining in which management and unions jointly bargain for wages and working conditions. Here the power of government through the National Labor Relations Board has some influence on what can or cannot be done.

<table>
<tr><td>

TYPES OF PARTICIPATIVE
MANAGEMENT

</td><td>

Participative management is not a one-way relationship; it requires a joint response from the engineer manager and the engineer. Both must recognize that participative management is a joint relationship. Each cannot have separate goals and objectives. To the engineer manager, this does not imply that his

</td></tr>
</table>

engineer manager responsibilities are reduced. He is still held accountable for performance, the actions of his group, and getting the work done. He has merely brought others to assist him in the decision-making process.

As an aid to understanding various kinds of participative programs, Beach has developed an interesting classification made up of two broad categories: (1) informal and semi informal methods and (2) formal programs.[1] Below is a discussion of these catagories:

1. *Informal and semiformal methods* involve relationships between a manager and his subordinates. These relationships tend to take place at regular intervals and primarily when a specific issue arises. At such times the manager may discuss the matter with one or more of his subordinates, or the department may have regularly scheduled meetings partly for information transfer and partly for group discussion and decision making. These methods include the following:

A. *Individual participation.* Individual participation may occur when the manager invites one of his subordinates into his office to obtain his thoughts on some particular problem, or perhaps while the manager is walking through the plant he may chat with one of the employees seeking information on ways to improve the operation. Sometimes the employee may need to take the initiative to speak with the supervisor. Such interchanges will take place only if there is mutual respect and trust. Both parties have to feel that they are working toward a common goal and that they can be frank in their interchange of viewpoints.

When the manager delegates some responsibilities and provides support to the employee, he is again providing an atmosphere for participation.

B. *Manager with group of subordinates.* A manager working with a group of subordinates finds a good platform for an exchange of ideas. Here an idea from one person may spark a thought in another. It may take place when the plant manager calls together his foremen to discuss matters related to safety or an impending strike. He seeks their ideas on how to best cope with the problem. In practice, the interchange can take various forms. It may be a matter of the manager's presenting to the group some decision which he has tentatively made. He elicits their comments

[1]Dale S. Beach, *Personnel,* 4th ed. (New York: Macmillan, 1980), p. 509.

and suggestions. In such an atmosphere, as mentioned earlier, some ideas may be stifled, since employees and subordinates, not knowing the boss's position, may be reluctant to offer their suggestions. Or the manager may merely present a problem to the group and invite their suggestions. After a period of discussion in which as many thoughts as possible are drawn out, he may announce his decision. In all cases, however, the manager reserves the final decision-making authority to himself.

The term *consultative* has occasionally been used to indicate this type of relationship. It is possible to go even further in putting the decision-making process in the hands of the subordinates. The manager may give the problem to the group and ask them to come up with a suitable solution. He may go as far as to say that he will support whatever solution they develop as long as the whole group has been involved. In such cases, the manager does not play a dominating role in the discussion. This type of relationship has been known as *democratic* management.

2. *Formal programs* work only when an organizational structure has been prepared to accommodate them. In these, certain of the participants acquire formal positions and titles. Some formal programs are:

A. *Committees.* Committees are found in all organizations. They represent a group of people elected or appointed and created to perform some function or mission. Committees are developed for a number of reasons, among them being to secure a greater variety of ideas on the solution of a problem or the delaying of a decision of a manager through his assigning the matter to a committee. Or they may represent a true decision-making body such as the executive committee of a board of directors of a corporation.

Some committees, known as *ad hoc* committees, are created for some specific purpose and are disbanded upon the conclusion of that purpose. Others are permanent or standing committees, to which perhaps different individuals are appointed each year or in which membership has a certain life. One occasionally hears of committee management, which perhaps represents the idea of committees in its worst sense. Properly used, such a grouping of individuals can have its merits. The manager must be careful, however, in the committees that he creates as well as the ones on which he must perform.

B. *Junior board of executives.* These boards represent standing committees of managers who carry out studies and make recommendations to top mangement. This approach is sometimes known as *multiple management.* These boards are generally made up of middle management executives who are deemed to have abilities for further promotion. They serve for a definite period of time on these boards. During this period of time they are given many different types of problems and investigations to carry out for the organization. Upon reaching their decision, they make a recommendation to top management. A number of major corporations have used this style of management and declared it an unqualified success.

C. *Collective bargaining.* As mentioned earlier, this is a type of participation that is not wholly at the discretion of management. Unions, by following certain procedures and securing sufficient votes of employees, can be recognized by the National Labor Relations Board as the employee's representative. Once this has happened, management must follow certain rules on how to deal with this collective bargaining representative.

Unions have their elected leaders and these leaders meet from time to time with representatives from management in an endeavor to work out mutually agreeable means of operation within the confines of the contractual agreement. It is well for the manager to be aware of all the provisions of the contract, for considerable difficulties can arise if violations are incurred.

D. *Union–management cooperation committees.* These committees are formed by representatives of union and management to work toward the solution of problems affecting plant operation for the mutual benefit of the company and the employees. They deal with such areas as health, safety, and employee morale. Recommendations from these committees must be either accepted or rejected by management. Some managements are not overly happy about giving away some of their inherent rights through such committees.

E. *Suggestion plans.* Suggestion plans, also known as *suggestion systems,* are formalized systems established by an employer to encourage employees to submit ideas that will result in improvements for the business and the organization. Normally, payment of monetary rewards is a fundamental feature of these plans.

A well-run suggestion program can provide a number of real benefits to the organization. Among them are additional revenue and an upward means of communication. This is important because frequently managers two or three levels removed are not familiar with many of the problems faced by the employees. The program is a morale builder and, in addition, it provides the employee with some monetary reward. In practice, boxes known as suggestion boxes, are located around the plant. At any given time an employee may put in his suggestion. Normally, a committee meets on a periodic basis and reviews the suggestions. A well-publicized procedure is used to identify the dollar savings that the suggestion may provide and also ways to measure the value of intangible suggestions. Generally, a staff employee makes all the calculations so that the results can be presented to the committee;

then the committee makes its decision on awards or rejection of proposals. Considerable publicity is given to the awardee in order to encourage others to make suggestions at a later date.

F. *Worker participation–European style.* The central core of these systems consists of worker councils comprised of worker representatives elected by the employees in each plant or by the union. These councils articulate the employees' interests in their discussions with management. The system of organization varies from country to country, but in all cases, the primary objective is to provide a mechanism by which the workers can bring their views to the attention of management. Such organizations are mandated by law in many countries.

Effective participative management requires the engineer manager to develop a strategy. By having developed participative strategies, he will be able to have a system of approaches that are generally appropriate. Strategies that the engineer manager needs to consider are outlined below:

DEVELOPING A PARTICIPATIVE STRATEGY

1. ***Trust in the engineer manager.*** The engineers and subordinates must have trust in their engineer manager; otherwise they will deem it unlikely that participation will afford them any influence. Only when they are able to trust him are they likely to express their sincere views.

2. ***Belief in participation by engineer manager.*** For participation to be successful, the manager must believe that participative management will give positive results.

3. ***Confidence in subordinates.*** The engineer manager must believe that his subordinates' abilities can make a contribution to the firm's activities.

4. ***High degree of empathy.*** For participation to be effective, there is a need for the manager to understand and sympathize with his subordinates' positions.

5. ***Organization commitment.*** Top management must really believe that participation is not only desirable but also essential. A highly centralized organization tends to restrict management authority to upper levels of management. Therefore, the engineer manager has only limited opportunity to be involved in participation. The degree of participation depends on whether the organization is highly centralized or decentralized in terms of authority. The organizational structure established by management may discourage participation by giving middle and lower management little decision-making discretion, and thereby few opportunities to implement active participation. On the other hand, the attitude of top management and the delegation of authority to middle and lower management will provide many opportunities for participation. When the engineer manager is given full authority to make decisions, he generally wants to involve his subordinates and encourage active participation.

6. *Definite participation activities.* Not all the activities that the engineer manager faces are appropriate for participation. Many government legal regulations are of this type. Whether or not to conform is not a question here; nevertheless, participative management may be useful in deciding how to meet the legal requirements. Another is layoff during a recession. Again, participative management practice may be very productive in developing alternative ways to reduce costs or to implement such layoffs.

7. *Emergencies.* These do not allow for participation, but the engineer manager can still obtain views and suggestions from people for future applications. A good engineer manager can, through participation, develop contingency procedures in anticipation of future situations. Then when the situation does occur, the group has had experience and has considered the problem.

8. *Definite participation limits.* For the engineer manager practicing participative management, some activities are already determined; in other words, he has little or no control. For example, the amount of vacation time available to an employee has already been determined and is not a question. But determining how this vacation may be taken or how many may be gone at one time and still maintain production is an appropriate participative management activity. The engineer manager must provide broad areas of activities for participative management, but they must be defined by him to the group.

9. *Rewards.* If engineers and subordinates are to be enthusiastically involved in participation, there must be provided some significant rewards for all involved. One reward is increased influence for both the engineer manager and his subordinates. But such intangible rewards generally are not enough. Eventually, the manager must see increased production and the subordinate must experience increased salary or job satisfaction.

10. *Understanding of personal attitudes.* The engineer manager must be capable of understanding the personal reaction of his people to participation. Each one must be capable of adjusting to the attitude required in participative management.

Properly used, the above considerations can lead the engineer manager to develop an active successful strategy for participative management.

ACHIEVING PARTICIPATION THROUGH INVOLVEMENT   Participation is difficult to implement, but it has enormous potential. It is not an engineering management activity that can be short range; it needs to be thought of in the long term. Participation facilitates both planning and changes, because everyone feels a part and understands the change. In many cases, participative management can improve the engineer manager's plans and cause him to eliminate many poor ones. This process broadens the outlook of the engineer manager and helps all involved to develop a closer relationship.

Participation requires both mental and emotional involvement, plus a desire to contribute to the group objectives. To achieve true involvement, the engineer manager must recognize various conditions:

1. For participation to be effective, the manager must develop and provide a system that allows his subordinates to take part in the participative management process.

2. The engineer manager must maintain authority in the decision-making process, for he is still responsible for all decisions. To maintain this relationship, he must direct participation activities without losing control of the situation.

3. True participation involves mental and emotional involvement rather than physical skills. When the manager involves only the engineer's physical skills, it is not true participation.

4. Participation is more than mere approval of decisions already made by the engineer manager. Everyone must have an opportunity to contribute to the decision-making process.

5. For engineers and subordinates to be effective in participation, they must be trained in what participative management is and what its functions and purposes are in the engineering management process.

6. It is easier to change beliefs and opinions when subordinates are allowed to participate in group discussions.

7. Not all engineers and subordinates will react positively to participation, especially if the manager has an authoritarian attitude.

8. In participative management the manager must forego some of his formal authority in exchange for greater assurance that the conclusions reached in the participative process will be implemented more effectively.

9. Participative management provides the manager greater control over the acceptance of the decision by the group than authoritarian management does.

All engineer managers are faced with the problem of motivating employees to contribute to the participative management process. These contributions are essential if the full benefit of participative management is to be attained. A highly motivated employee in the participation process is one who works toward the objectives of the organization.

Participative management works very effectively with some kinds of activities, but it is very ineffective with others. To further confuse the issue, there are many managers who use an autocratic management style and have high productivity in their departments and there are many managers who use a participative management style and also have high productivity in their departments. Consequently, participative management must be used carefully if it is to be successful, and the manager must evaluate his situation to determine if participative management will be effective. Some of the factors that appear to indicate that participation may be successful are:

1. Better-educated individuals tend to be more responsive to participative management techniques than individuals who have had less education. This should indicate that engineers should be interested in participation.

2. An enthusiastic engineer manager who shows a great interest in his engineers and for participative management techniques will be more likely to succeed.

3. Participation is a middle-class value and grows out of the prior expectation of people being supervised. Thus, participation generally will be more effective with groups at higher economic levels.

4. The participative management concept must be presented by the manager so that the engineer thinks he is participating in the decision-making process. Even if he is actually participating but does not believe he is, participation will be ineffective.

5. Research has demonstrated that subordinates who have strong independence needs react more favorably toward the opportunity to participate in decision-making activities than those who have low independence needs and who score high on the authoritarian measuring scale. This indicates that engineers should work well with participation.

6. A high degree of delegation on the part of the engineer manager contributes to an environment for participation.

For participative management to be effective, the manager must have enthusiasm for participation. Further, the entire organization, must believe in and practice this philosophy, from top management to first-line supervisor.

SOME COMMON ERRORS IN PARTICIPATIVE MANAGEMENT

A careful scrutiny of past and present participative management practices will show mistakes and possible improvements in future operations. Here are a few of the problems that engineer managers are likely to encounter when they are engaged in participative management in their departments:

1. *Failure to recognize their own biases.* At times a deeply felt desire to achieve one of his particular goals, or a boss's goal, blinds the engineer manager to the actual conditions. When this occurs, the manager concentrates on justifying or obtaining evidence supporting what he believes instead of starting with the facts and asking members of the group what their ideas are. This may happen when the manager puts together an analysis that shows proof to his group that what he wants to achieve can occur. But when the group analysis arrives at a different conclusion that he fails to accept, participative management is not present.

2. *Incomplete facts.* Another form of logical error occurs when key facts are omitted. Some factor outside the group may have an important bearing on the question. It is the manager's responsibility to be sure that the information is made available to the group.

3. *Poor information.* Often decisions are made by the group on what is presumed to be accurate data. This occurs when a staff member of another department provides to the group information that is either inaccurate or incomplete because of lack of time or the cost of obtaining such information. When a decision is weighted in favor of inaccurate data or misinformation, it can lead the group to false conclusions. Because the firms operate in a highly dynamic environment, even information that is accurate when obtained may be misleading when the group acts on the problems.

4. *Information based on one event.* It is false logic to establish targets based on a single event instead of considering the data as a point within a range. This occurs when the event represents a period of time when the operation was either very profitable or very unprofitable. An analysis based on one event fails to recognize the standard variations that inevitably occur.

For the engineer manager starting out to install participative management in his department, the following questions arise: How and where do I start? What steps are involved? A workable procedure for the engineer manager is set forth below:

IMPLEMENTING PARTICIPATIVE MANAGEMENT

1. Identify the problems of the organization for the coming year. Some typical areas in which problems may develop are:
   A. Technological development
   B. Productivity changes
   C. Equipment improvement
   D. Design product changes
   E. Product innovations

2. Participative management must involve each person in the process. The manager should take the following actions to ensure each person will be involved:
   A. Request all to study and make notes on the problems before the group meets with him or before they meet individually with him.
   B. Before the meeting begins list some solutions to the problems and look for any effect on established policies.
   C. During the conference obtain suggestions and comments from the group before offering suggestions.
   D. Appoint someone to prepare a draft of solutions developed during the meeting to circulate to all attendees. Request comments.
   E. Have the draft revised and recirculated.

3. Call meetings of the group periodically to review goals and progress toward them. The results of group decisions should be reviewed from time to time. The manager should:

    A. Make sure he is doing his part.

    B. Use the jointly agreed-upon decisions as a tool for coaching, training, and improving performance of the group.

    C. Reinforce good results with rewards.

4. Call upon the group to evaluate results. Determine causes of failure or success.

5. Motivate members of the group to greater participation by showing the effects of their participation.

Simply setting forth an outline of this type may give the misleading impression that participative management calls for no more than following a prescribed procedure. Actually, this overlooks the factors of time, judgment, and individual situations, all of which are vital parts in making participative management effective. It is very important for the engineer manager to realize that everyone clings to his own expectations, ideas, attitudes, and values. No participative management process can sweep these aside or ignore them. In the participative management process, the engineer manager must allow for these and go forward at a rate determined by the group.

For participative management to be effective, it must have the full endorsement of top management. Unless top management and key people in the organization know, accept, and use participation, the engineer manager will have difficulty in using it effectively. If either top management or the engineer manager resists giving up personal control, trying to develop a participative management program will be difficult.

One of the important reasons for involving top management in the process is to engage them in establishing common objectives that will direct and guide the group's participative management process. Top management must establish goals for the entire organization before a department can establish goals that will coordinate with them. By doing this, top management can have all departments working together toward the same goals. In this manner, ideas from throughout the organization can be brought to bear on achieving the corporate goals.

CONSTRAINTS AND
BARRIERS TO
PARTICIPATIVE
MANAGEMENT

A common barrier to effective participative management is the manager who has an authoritarian personality. Such a manager finds the very idea of consultation with subordinates an anathema. He cannot bring himself to accept suggestions from people beneath him in the hierarchy of the organization. Another barrier to real participation is the executive who likes the trappings of participation but does not want to share the decision-making responsibilities. These managers develop what they call a "sense" of participation with their employees. They encourage them to work more enthusiastically but do little else to develop a team relationship.

Subordinates are sometimes manipulated through meetings. The meetings are called ostensibly for joint consultation and problem solving, but the manager arrives at the meeting with his mind made up. He uses the meeting to secure the support of others. This is not good management because it hurts the manager's credibility with his employees. If he wants to sell them on some particular pet scheme of his, he should attack it head-on with his best efforts of selling and persuasion and not hide it under the guise of participation.

Another barrier to full participation comes from the fact that the supervisor is part of the participative team. Many subordinates find it difficult to speak freely in the presence of the boss. They recognize that he controls their future by way of raises and promotions; therefore, they are reluctant to come forth with ideas that might actually be quite helpful in the decision-making process. Some of them may well know the boss's pet ideas and get on the bandwagon to support them, even though they may recognize great difficulty in implementation. Obviously, it takes a special effort on the engineer manager's part to break down this particular barrier.

The manager can also help by not expressing his own views on the matters under discussion until those of others have been heard. In this manner, he may be able to obtain more genuine uncolored contributions from his subordinates.

Another subtle barrier results from pressures to conform to the common beliefs of the group and of the larger organization. It occurs in both formal and informal groupings of departments, divisions, and companies. This pressure for conformity tends to emerge and develop in order to protect the stability and security of the group or the formal organizational unit.

SUMMARY

Participation is the term used to designate the process by which people contribute ideas toward the solution of problems affecting the organization and their jobs. It allows people to exercise some degree of influence in the decision-making process. It is appropriate at all levels of the organizational hierarchy from first-line supervisor to top management.

There are two major types of participation: informal and semiformal methods, and formal programs. Coming under the informal and semiformal types are individual meetings between manager and subordinate and periodic meetings between manager and a group of his subordinates. The more formal programs involve committees, junior boards of executives, collective bargaining, union–management cooperation committees, suggestion plans, and worker participation–European style.

The relationship between a manager and his group of subordinates is probably most frequently recognized as participative management. In this environment the subordinates are brought into the decision-making process, creating perhaps a better solution to the problems facing the group, and certainly a better acceptance.

Among the formal programs, perhaps committees and suggestion plans are most common. Committees are either of an *ad hoc* nature created for a specific purpose or of a more permanent nature. In either event, the membership tends to change over a period of time. Suggestion plans or suggestion systems are a means by which ideas can be solicited from employees which can lead to improvements in operations and cost savings. Normally, financial rewards are made to those whose suggestions are accepted.

It is important to develop a participative strategy. In this strategy there must be an element of trust of the engineer manager by the subordinates, and the engineer manager must himself believe in participation. There must also be an organizational commitment. Certain limits of participation must be defined and rewards be given.

A number of common errors in participative management include the failure of managers to recognize their own biases and the group's working with incomplete facts, poor information, and information based on only one event.

It is important to involve all members of the group in order that participatio may be effective. The engineer manager must maintain his authority, even though his subordinates are brought into the decision-making process. True participation involves mental and emotional involvement. To be effective, engineers and subordinates must frequently be trained to work in a participative environment.

There are certain factors which appear to indicate that participation may be successful, including the educational level of the individuals, the enthusiasm of the manager, the economic level of the group, and the degree of delegation that the manager has practiced.

There are a number of barriers to the effective operation of participation. A major barrier is the personality and management style of the engineer manager. Seeking consultation with his subordinates may be an anathema to him. As a result, he may pretend to be interested in the trappings of participation without genuinely working in concert with his employees. Further difficulties occur because the subordinates are frequently unwilling to state their views in the presence of the boss.

To implement participative management, the manager must identify the problems of his department. He must involve each person in the problem-solving process and periodically call meetings to review the goals and progress toward achieving them. He must call on the group to evaluate the results and use these results to motivate them toward greater participation in the future.

Important Terms

Collective Bargaining: The bargaining that takes place between union and management representatives to determine salaries and working conditions for union employees.

Individual Participation: An informal relationship between the manager and his subordinate leading to an interchange of information and advice.

Participation: A term used to designate the process by which people contribute ideas toward the solution of problems affecting the organization and their jobs.

Participative Management: A management technique involving subordinates and supervisors in the participation process.

Suggestion Plans: A formalized system established by an employer to encourage employees to submit ideas which will result in improvement for the business and the organization.

For Discussion

1. Define participation.
2. Is participation appropriate for all levels of the organizational hierarchy? Why or why not?
3. Name three barriers to participation.
4. Name the two major types of participation.
5. How does participation work between the manager and the individual?
6. Why do committees represent a form of participative management?
7. What are suggestion plans and of what value are they to the corporate operation?
8. What is the function of junior boards of executives?
9. Why would collective bargaining be called a participative process?
10. Name five elements of a participative strategy.
11. Why is involvement of importance in the participative process? What conditions are important in bringing about such an involvement?
12. Name four common errors in participative management.

13. Name four steps for the implementation of participative management and explain how they work.
14. Give some examples of situations in which participative management might not be effective.
15. Why has committee management come under so much criticism?

Case 9-1

The Newly Promoted Engineer Manager

Before Rex Smith's official appointment to supervisor of the engineering department he had been working as an engineer for ten years. His work as an engineer had consistently been superior.

Rex's co-workers all wished him well on the new job and for the first month most of them were cooperative and helpful while Rex was adjusting to the new role.

But before long an incident occurred that portended trouble. About 4:45 he ran onto two of his engineers coming out of the locker room with their coats and hats. "Say, guys, you should not be leaving so soon; it's only 15 minutes till five" said Rex. "Go on back to your work and I'll forget I saw you leaving." "Come off it, Rex. You used to slip out early yourself when you had a hard day. Just because you've got a little rank now, don't think you can come down on us." To this Rex replied, "Things are different now. Both of you get back on the job or I'll have to take disciplinary action." Both said nothing more and both returned to their work.

From that time on Rex began to have problems as a supervisor. The group seemed to forget how to do the simplest things; they showed no imagination. Drawings were prepared improperly and complaints increased. Everything seemed to go wrong and by the end of the month Rex's department was behind schedule, work was being returned, and drawings were incomplete.

1. How should Rex have handled the incident?
2. What would you suggest Rex do to get the department performance back on track?

Case 9-2

Fun and Games in the Group

Lately, the engineers seemed to be doing everything to hamper Rex Smith in his new position as supervisor of the engineering department. None of them had anything personally against him, but all of them considered it a game to pit their wits against his. They set up signals to inform them when he left his office so everyone would appear to be working hard. They tried to draw him into doing some of their work while they stood around. They complained constantly and without justification about the lack of data and information on the projects.

At breaks and lunch time they bragged about the latest trick they had pulled on Rex and planned new ways to harass him. They openly ridiculed the department and the company. All of this seemed to be great sport and each had a lot of fun. Actually, these actions were becoming a habit with the group and the constant complaining was greatly reducing their productivity and morale.

Now Rex is determined to change this attitude and win the group over to working for the department instead of against it. He knows that if this can be done, he will have a top-notch group. Each knows his area and they have a very good team spirit. The problem is getting them to use their knowledge and efforts constructively.

1. What suggestions would you make to Rex?
2. How should they be implemented?

For Further Reading

ARGYRIS, C., *Integrating the Individual and the Organization.* New York: John Wiley & Sons, Inc., 1964.

BRECK, EDWARD, "Don't Hand Over Just Work, But Responsibility," *Business Administration* (May 1977), 28–30.

CHANEY, FREDERICK B., and KENNETH S. TEEL, "Participative Management: A Practical Experience," *Personnel* (November–December 1972), 8–19.

DUBIN, R., *Human Relations in Administration,* 4th ed. Englewood Cliffs, N.J.: Prentice-Hall, Inc., 1974.

FIEDLER, F. E., *Leader Attitudes and Group Effectiveness.* Urbana: University of Illinois Press, 1958.

FURLONG, JAMES C., *Labor in the Boardroom.* Princeton, N.J.: Dow Jones Books, 1977.

LIKERT, RENSIS, *The Human Organization.* New York: McGraw-Hill Company, 1967.

———, *New Patterns of Management.* New York: McGraw-Hill Book Company, 1961.

MARROW, ALFRED J., et. al., *Management by Participation.* New York: Harper & Row Publishers, Inc., 1967.

MCGREGOR, DOUGLAS, *The Human Side of Enterprise.* New York: McGraw-Hill Book Company, 1967.

MILES, RAYMOND E., "Human Relations or Human Resources?" *Harvard Business Review* (July–August 1965), 148–63.

WILLIAMS, ERVIN, ed., *Participative Management: Concepts, Theory, and Implementation.* Atlanta: Georgia State University, 1976.

CONTROL TECHNIQUES FOR THE ENGINEER MANAGER

The engineer manager is responsible for the measurement of accomplishments against plans and the correction of deviations to assure attainment of objectives. It is necessary to measure progress in order to uncover deviations from plans and to take "corrective action." To measure performance, the engineer manager must be familiar with return on investment, control charts, break-even analysis, marginal costs, and average variable costs.

☐ Understand the criteria for an adequate control process and the importance of feedback.

☐ Learn the characteristics of an effective control system.

☐ Gain insight into accounting as a control device.

☐ Understand the construction of income statements and balance sheets.

☐ Understand the various budgets that are effective in control.

☐ Become familiar with the construction and use of a PERT chart.

EXECUTIVE COMMENT

JOHN M. LANG
Public Works Administrator
City of Portland, Oregon

John M. Lang joined the Office of Public Works Administration of the city of Portland in 1975 after several years as director of Public Works for the city of Walla Walla, Washington. After four years as chief civil engineer he was appointed Public Works administrator in 1979. His responsibilities encompass engineering design for all city sewerage and transportation improvements as well as maintenance and operations.

Control Techniques

It has been said that control techniques are just as necessary for an engineer manager as a guidance system is for a military missile. Certainly both the engineer manager and the missile have goals or targets to achieve and to do so they must have sufficient feedback for course corrections in order to reach the target accurately. Most engineer managers in public agencies utilize several different control techniques to ensure that their goals and objectives are met and that their work is responsive to public needs.

Basic to control techniques in public agencies is the annual budget which identifies the project and activities to be accomplished and which appropriates the necessary funds and personnel for that work. Once the budget has been approved, an accounting system should be used to monitor and control the expenditures during the year. Included in the system should be a procedure for forecasting the year's monthly expenditures and making monthly reports on actual expenditures experienced.

In addition to budgetary controls, other techniques are used by the engineer manager. Work performance standards, project control systems, and a complaint monitoring system all provide invaluable control information. Properly used, these techniques can be implemented by individual employees and work groups to set work objectives, to receive feedback on their performance, and to identify needed changes for meeting the objectives. These control techniques become a way of increasing employee productivity as well as supplying the engineer manager with needed information about the agency's performance.

Public agencies exist to provide services for area residents. A good control system can guarantee the engineer manager that his budget expenditures and employees' work are efficiently and effectively providing those needed services to the residents.

Once plans have been implemented, the engineer manager is responsible for using control techniques to measure progress toward planned objectives, identify deviation from plans, and take corrective action. Therefore, the control process must involve more than measurements. It must include changing methods of operations so as to meet original plans or, if necessary, revise the plans.

The more complete and clear the plans are, the more effective the controls can be. It is impossible to design a control system without a careful study of the plans, especially of how well the plans are formulated.

Control systems also require that there be an organizational structure in place. Control of activities operates through people. It is not possible to know where the responsibility for deviations and needed actions will lie unless the organizational responsibility is clearly assigned. Therefore, a major prerequisite of control is the existence of an organizational structure. In establishing the control techniques, the engineer manager must not only tailor them to the plans that have been developed, but he must also consider the method by which the resources were organized, including the needs of individual personalities.

The engineer manager when he worked as an engineer considered control devices in the design of processes. In fact, he designed systems with feedback which, in many cases, were forward control systems. Most controls that the engineer developed were physical, but now as an engineer manager, many of these are financial in nature. Therefore, he must become familiar with accounting control devices, such as budgets, profit and loss statements, and investment criteria.

The control process has been defined by Koontz and his colleagues as made up of three steps:

1. Establishing standards
2. Measuring performance against these standards
3. Correcting deviations from standards and plans[1]

[1]H. Koontz, C. O'Donnell, and H. Weihrich, *Management* (New York: McGraw-Hill, 1980), p. 722.

Since plans are the yardsticks against which controls must be devised, the first step must be to establish plans. Most plans that the engineer manager develops are detailed and complex, and therefore special standards for measurement must be established. These standards allow the engineer manager to measure performance with the primary purpose of providing himself with information on the progress in executing but without having to oversee each and every step of the plan's implementation. Standards may be of many kinds, but the best are verifiable goals or objectives. The end results for which people are responsible are the best measure of plan achievement.

Standards should be drawn up so that subordinates know exactly what they should be doing and so that appraisal of actual versus expected performance can be made. It is easier to do something like this in labor-intensive operations producing mass items than it is in a group of engineers. In the latter case, judgments have to be more subjective. But they can be quantified to a degree by setting up time schedules for the accomplishment of tasks.

If possible, such measurements of performance should be made on a forward-looking basis (sometimes called *feedforward*). In the case of the production operation, a forward-looking manager might predict that because of some needed machine maintenance the productivity of the unit is going to decrease. In the case of the group of engineers, strikes anticipated in the future might have an effect on the ability to secure equipment for meeting the construction schedules.

It is implicit within the definition of control that deviations can be expected. If the manager has set up appropriate standards and appropriate means of measuring performance, then it is easy to implement the third step of corrective action. Deviations may be either positive or negative. The manager should have as much concern for analyzing why there is a positive deviation as for a negative one. In either case, he should be prepared to redraw his plans or modify his goals. In support of this, he may have to reorganize his function or reassign duties. The correction of deviations is at the heart of the control process and should command a significant portion of the engineer manager's attention.

CRITERIA FOR ADEQUATE CONTROL PROCESS

The engineer is familiar with the mechanical cybernetic system as exemplified by the steam engine governor. In order to control an engine's speed, balls are mounted so that the diameter of the circle in which they travel will vary with speed. As speed increases, the diameter of the circle increases through centrifugal force acting on the balls. This causes movement of the anchor points closing the steam valve. This in turn reduces the speed of the engine. As the speed reduces, the reverse actions occur.

Norbert Wiener pointed out that the communication, or information transfer, and control occur in the functioning of many other systems including physical, biological, and social.[2] They all have in common the need for an information

[2]Norbert Wiener, *Cybernetics: Control and Communications in the Animal and the Machine* (New York: John Wiley, 1948).

feedback in the form of written or oral messages, electrical impulses, or chemical reactions. In other words, systems to be effective must have information on deviations from standards fed back so that corrective action can be taken.

Management control, then, can only take place if there is an effective feedback system. As management finds deviation from the goals or objectives as shown by established standards, management can make the necessary corrections in order to continue toward the original goals or objectives or to modify them accordingly (Figure 10–1). In order to make full utilization of the feedback principle, the control process should be designed with the following characteristics:

1. *Appropriate to what is being measured.* Since every process being controlled is different, the manager must initially determine clearly what he wants to measure in his particular system. In production it might be costs, in sales it might be dollars of sales per salesman within a given period of time, or in an engineering group it might be completion of the design of a new plant facility by a given time.

2. *Abilities of the people involved in the control process considered.* The feedback that the manager might expect from nontechnical hourly employees operating in the plant would be different from what he might expect from highly trained engineers working on the design and construction of a major new facility. The data that he would expect, the means of transmittal, and the time frame would vary between the two cases.

3. *Appropriate benefits versus cost.* As the control system is developed, the engineer manager must assure himself that the cost of this operation does not exceed the benefits that are to be obtained. It is important that he analyze how the information will be used, its quantity, and how much it costs to obtain it. He must also consider whether or not the paperwork involved in control will significantly slow down the constructive work the engineer might perform.

4. *Workable controls.* Because the organization must operate in a dynamic economy, controls must be workable. During these changing conditions the manager must consider the effect of the environment on the process he is trying to control and make the necessary adjustments. These adjustments may be the result of changing patterns of sales competition, changing labor rates, or inflationary impact on materials. This may necessitate redoing budgets or perhaps developing a model that will allow the budget to be altered on a continuous basis depending on such changes.

5. *Impartial to all involved.* Any control process developed will involve intangible factors requiring subjective judgment. The engineer manager should keep these judgments to a minimum in order to avoid the influence of personalities. If at all possible, the control system should be designed to work with quantitative information.

6. *Realistic*. The manager must ensure that his controls are realistic. For example, in an assembly operation some items such as screws and bolts are not worth maintaining controls on. But high-priced items need close control because of costs as well as availability.

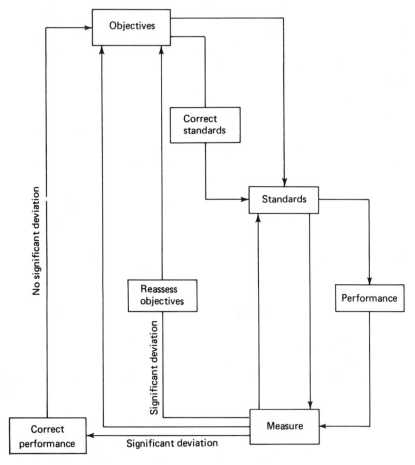

FIG. 10-1. The Control Process. [Source: Robert J. Mockler, "The Control Process," The *Management Control Process* (New York: Appleton-Century-Crofts, 1972), p. 3.]

MAKING THE CONTROL FORWARD LOOKING   Because of the time lag in the management control process, there is a need for a forward-looking or feedforward control system. Feedback as previously described is really an historical type of control. An effective manager needs a system that will predict what is going to happen so that he can take corrective action before it occurs.

One common way that managers can have forward-looking control is through repeated updating of forecasts. When the manager sees deviations from the original forecasts, he can make decisions toward achieving the original goals. A simple example that occurs over and over again in the corporation deals with sales forecasts. When management sees that the revised forecasts indicate that their sales will be dropping and that they will miss their anticipated goals, management can move to correct this by increasing the sales promotion or introducing new products.

This technique is also very important in the financial area. A company must have adequate cash in order to operate its business. Forecasts of cash availability help the financial manager to make necessary arrangements to correct the difficulties that he sees ahead.

The engineer has an excellent technique for forward-looking control in network planning as exemplified by PERT (Program Evaluation and Review Technique). We will discuss this in detail later in this chapter. We might say, however, that it does represent a planning technique from which deviations from standards can be seen and correction made.

In developing an effective control system, a number of characteristics should be considered important. The system should be:

ELEMENTS OF AN EFFECTIVE CONTROL PROCESS

1. *Timely*. If controls are going to be of value, they must provide information in a timely manner. In a production process it is not profitable to have a quality control system that shows a deviation from standards after 20,000 gallons of high-priced materials have been produced below the quality standard. Similarly, accounting systems that produce financial control data several weeks after the close of the month are not acceptable means of timely financial control. Therefore, any control systems, whether they deal with physical or financial matters, should be structured so that the manager receives the information in time to make corrective action.

2. *Economical*. A desirable control system should be economical to operate. The benefits to be achieved must bear the proper relationships to the costs that will be incurred.

3. *Understandable*. The control should be easy to understand by both managers and operators. This at times works against using some of the more sophisticated control devices. The control system should be acceptable to those involved.

4. *Flexible*. The environment, both internal and external, is continually changing within an organization. The control system must have the ability to conform to these changing environmental conditions. Internally, there may be changes in the process which provide the capability for more efficient operations, to which the control system must respond. Externally, competition or changing economic conditions may well affect the standard of performance.

5. *Appropriately located.* Control should be located throughout the organization in order to most effectively bring deviations affecting the meeting of overall goals or objectives to the attention of managers. A manageable system of controls must be broken down into manageable elements that can integrate together and work toward common goals. In this placement it is well to look for operations in which exceptions to standard performance can be most effectively noticed. In managing by exception the manager does not find his time consumed by reviewing mountains of data in which no significant deviations exist. One thing the manager has to guard against is being swamped with too much information.

6. *Tailored individually.* Controls must be tailored to the personalities of the managers. Some managers feel more comfortable when reviewing financial statistics. They find that they communicate the kind of information they need to know in order to make corrections. Other managers feel more comfortable with production information dealing with process and product data and with matters related to the personnel involved. Some managers like their data in complex table form or in the form of computer printouts. Others like charts and written communications.

7. *Correctable.* The essence of any control system is to be able to provide corrective action. Therefore, all that is gone before is useless unless it ultimately ends up in information that can lead to an in-course correction. Controls are only justifiable if in their application they permit the adjustment of course so that the goals or objectives of the organization may be achieved.

CONTROL TECHNIQUES

There are many control techniques that the engineer manager will use in his operation, but there are two major ones that he will use most frequently: (1) budgeting and its supporting accounting techniques and (2) time event and network analysis, known as Program Evaluation and Review Technique (PERT). Since these techniques are valuable to either the engineer manager working in production or in engineering, they will be discussed below in some detail.

ACCOUNTING CONTROL TECHNIQUES

One of the most common sources of control information is accounting data. The engineer manager must become familiar with the accounting and budgeting methods that will be required for his department's operation. Accounting information is derived from the income statement and balance sheet, which help to provide an overall view of the organization's operation during the past accounting period. The accounting system provides the basis for the financial statements, with the objective being to serve the needs of the manager in his decision-making responsibilities.

The Income Statement

The income statement may be referred to as a statement of income, a profit and loss statement, or an income and expense statement. It is dynamic in nature, for it shows the results of the operations for a given period of time. The basic accounting equation for the income statement is: revenues minus expenses equals profits. Revenues are derived from the sales of goods and services. The expenses are the cost of goods sold and the sales, R&D, and administrative expenses. Some of the most common expenses found in an income statement are for the following:

1. Direct labor costs
2. Supervisory salaries
3. Office salaries
4. Research and development
5. Raw material
6. Real estate taxes
7. Insurance
8. Maintenance
9. Transportation
10. Selling
11. Utilities

These expenses are divided into direct and indirect costs. Direct costs are expenses such as labor and materials that can be directly related to a particular job, production of a product, or a project. Indirect costs include insurance, taxes, supervisory and office salaries and selling and corporate expenses that cannot be assigned specifically to a product or a project. In a product operation a plant production cost is determined by including direct and indirect plant costs. To these are added all other corporate administrative and selling expenses before subtracting from sales.

Revenues are a result of providing products and services to the firm's customers. In the income statements are listed all revenues minus returns and allowances and cash discounts to arrive at the net sales for the period. Cost of goods sold is computed and subtracted from the net sales to obtain the gross margin. Selling expenses, R&D, and administrative overhead are then subtracted from the gross margin to obtain the net margin (Table 10–1).

The *gross margin* measures the difference between net revenue from goods and services and their plant level cost. The gross margin is significant to the engineer manager for comparison of gross margin as percent of sales with other similar operations and with last year's operations.

In calculating cost of goods sold, the cost of current production is corrected for the change in finished and in-process inventory value over the period covered

```
                          TABLE 10-1

                         ABC Company
                       Income Statement
              For the Year Ending December 31, 19xx
```

Net Sales			
Gross Sales		$4,450,000	
Less Refunds and Allowances		(100,000)	$4,350,000
Cost of Goods Sold			
Material		$2,500,000	
Direct Labor		700,000	
Plant Overhead		220,000	
		$3,420,000	
Gross Margin			$930,000
Selling and Administrative Expense			
Selling		$ 400,000	
R&D		200,000	
Administrative		100,000	
		$ 700,000	
Net Margin			$230,000
Taxes			$115,000
Net Profit			$115,000

by the statement. An increase in ending inventory value reduces the cost of goods sold.

Cost of Goods Manufactured. The major item in cost of goods manufactured is the cost of the raw material. This is the cost of the material that will be upgraded into a finished product. In the case of a chemical company, it might mean a change from one chemical into a chemical of an entirely different character. In the case of a tank company, it might mean the conversion of steel plate into a finished product that would handle 10,000 gallons of liquid product.

Normally, raw materials are inventoried, meaning that they have been bought in quantities in excess of the current usage. In producing the income statement it is necessary to correct the cost of raw materials purchased during that month by a formula such as: *Raw material cost equals value of beginning raw material inventory plus purchases of additional raw materials minus value of ending raw material inventory.* The manager must make a judicious balance between assuring himself of adequate raw material availability versus the cost of maintaining large quantities of these materials on hand.

The main element in converting the raw material into a finished product may be labor costs. Generally, these costs can be allocated against a product or a production line and therefore appear in the income statement as direct labor. Such labor might include the operation of equipment as well as the unloading and handling of raw materials and the handling and loading of finished products. All companies do not record these items of costs alike, however. Some include the second and third items of labor as overhead cost, feeling that they cannot allocate them precisely against a given product line. It can be seen that if they are incorporated as direct labor cost, the direct labor figure may be distorted a bit, since some costs may be allocated to raw materials not yet used or to finished products not yet sold.

For a reasonably accurate allocation of direct labor cost, time sheets are frequently used. But even this does not allow an exact interpretation of cost because a certain portion of the raw material will not arrive at the finished product stage during the course of the month. At the end of the month it will be recorded as work-in-process inventory. Much judgment must be used as to how to value the work-in-process inventory. Some companies value it at only the raw material value. Others value it at raw material cost plus labor, and still others place a judgmental value on it in relation to its percentage of completion toward a finished product. The latter is the most generally accepted method. It can be dangerous to inflate the work-in-process inventory by placing too high a value on it because auditing at some future date may require a reduction in value. It is tempting, however, for the plant manager to overvalue because raising the value of the work-in-process inventory will reduce his cost of goods sold and improve his gross margin.

Heat, light, and power are normally required in the production of goods. When these items can be allocated against the production operation, they should be. But if this is impossible, they should be included in the overhead cost. Maintenance of the operating units is sometimes allocated directly to the product line if adequate data are available. If not, this cost may show up in the overhead.

A number of items cannot be allocated to a specific product line; oil, gasoline, cleaning, and paint are examples. These costs normally end up in the factory or plant overhead. Other costs that end up here include the plant manager's salary as well as the salaries of the superintendent, personnel manager, secretaries, clerks, accounting department, safety department, procurement department, and personnel department and insurance and taxes. Since these items cannot be specifically allocated against a certain segment of the plant, they are accumulated as a total indirect cost. If one wants to account for total cost of goods sold by product line, frequently he takes this total overhead cost and prorates it to the various product lines based on direct labor.

There is an item that appears on the plant level production cost sheet which is of a noncash nature. This is the item of depreciation. The Internal Revenue Service (IRS) allows a company to expense a certain percent of its capital investment over a specific period of time. The so-called depreciation rates are established by the financial manager for the concern, frequently in consultation with his auditing firm and with the Internal Revenue Service. Depreciation alerts the management of the company to the fact that at a future date there will be a cost to replace equipment that is wearing out. The amount of allowable depreciation is not intended to accumulate cash to replace the equipment. Therefore, the corporation must make plans to borrow or to provide a future source of income to replace equipment.

Selling, R&D, and Administrative Expense. Selling expense includes all those items important to the marketing of a product, including advertising and direct sales. In the multidivision corporation the total selling and advertising expense may be accumulated and then reallocated back to the various plants. In the consolidated balance sheets and income statements for the company it is shown as a single item.

Research and development costs are normally charged against income as they occur. These costs include the expense of operating laboratories, pilot plants, and new product engineering. Research and development costs may create intangible assets such as patents, trademarks, leases, perpetual franchises, and licenses. These assets that have a fixed life, for example, patents and trademarks, are charged against income and are prorated over their life span.

Administrative expense covers all those other costs at the corporate level including the president's salary, all of the staff salaries represented by public relations, human resource departments, traffic and transportation, and marketing. These generally represent only a few percent of sales and are allocated back to the division on a sales basis.

Importance of the Income Statement

The engineer manager in the operating segment of the corporation will find the income statement a vital portion of his life. Each month he will be concerned with the various costs contributing to the cost of goods made and to the cost of goods sold and he will be intensely interested in whether the sales value allows him to show a profit or a loss. He will also be concerned that his control systems show deviations from projections of profit or loss, and why these have come about.

The engineer manager who is involved in the engineering department will be less concerned with all of the production costs previously described, at least on a monthly basis. He will, however, be concerned with the creation of pro forma income statements in connection with new processes and new product lines. He will be involved in forecasting profits over a number of years that will result from such new ventures. He will be incorporating these into proposals which he will present to management to justify the capital expenditure required to provide facilities for such a new product line. At a later date he will also be concerned in following up to determine the deviation from his forecast and the actual operating profit.

His activities may, however, bring him into a close relationship with the day-to-day profit and loss. If he is concerned with process or industrial engineering studies, he will be involved in the determination of how to improve the operating efficiency of production lines. Here he will be studying past performance and forecasting how to improve performance through changes in operation. Once more he will be preparing proposals and forecasts, but they will be on the shorter-term basis than the ones for new product lines.

Balance Sheet

Another important accounting tool is the balance sheet (Table 10–2). This is a listing of the firm's resources and the creditors and owners of these resources. The firm's resources are called *assets*. The owners who have a claim against the firm's assets are referred to as having an *equity* in the firm. This owners' interest in the assets is called *owners' equity* or sometimes *shareholders' equity*. Creditors' interests in the assets are called *liabilities*.

The balance sheet is a statement of the financial condition of the firm at a given date and it reflects the results of all activities recorded since the firm was organized. Therefore, the balance sheet is a cumulative report. The status of the assets, liabilities, and equity of the corporation at any given time is revealed. The balance sheet provides information at a particular time but does not give information on what changes occurred or why changes occurred. The accounting equation for the balance sheet is: *assets equal liabilities plus owners' equity*. A net profit (after taxes) as shown on the income statement adds to the owners' equity on the balance sheet through the returned earnings shown on the balance sheet.

In the balance sheet the assets, liabilities, and owners' equity are listed separately and are not reduced by offsetting each other, even though specific assets may be used to secure a liability. The reasoning is that liabilities are paid in cash and assume pledged assets only if the firm defaults and the owners have an undivided interest in the total assets. Thus, the liabilities and owners' equity are the rights to the total assets, not to any particular assets. The assets listed are current, long-term, and other; liabilities are current, long-term, deferred revenues, and other.

Current assets include accounts receivable, inventories, and prepaid accounts (assets that will be converted into cash within the coming year). *Long-term assets* are equipment, buildings, and land used in conducting the operations that will *not* be converted into cash during the coming year; these are used in the operation over a relatively long period of time. These are not expected to be sold and converted into cash. Buildings and equipment that have a limited life are shown at cost less accumulated depreciation to give net book value. Because land has unlimited useful life, it is not depreciated.

Assets are not adjusted to current replacement costs or to net realizable values. These assets are useful in performing the work of the firm, but the cost is recovered gradually over time as they are used in the production process. However, when equipment is to be replaced, the current replacement cost and the

TABLE 10-2

ABC COMPANY
Balance Sheet
December 31, 19xx

Assets			Liabilities		
Current Assets			**Current Liabilities**		
Cash	$ 200,000		Accounts Payable	$1,200,000	
Accounts Receivable	1,000,000		Tax Liability	100,000	
Inventories	600,000		Accrued Expense		
Prepaid Expenses	50,000		Payable	50,000	
		$1,850,000			$1,350,000
Fixed Assets			**Other Liabilities**		
Land, Building, and			Mortgage	$ 500,000	
Equipment	$3,000,000				$ 500,000
Less Accumulated					
Depreciation	1,000,000				
		$2,000,000			
Other Assets			**Stockholders' Equity**		
Investment	$ 100,000		Common Stock	$1,500,000	
Patents, Goodwill	100,000		Retained Earnings	700,000	
		$ 200,000			$2,200,000
Total Assets		$4,050,000	Total Liabilities		$4,050,000

value of trade-in of present equipment is considered in the decision. For tax purposes, the IRS requires trade-ins on new equipment to be valued as follows:

Old asset's cost	$5,000	
Old asset's accumulated depreciation	(3,000)	
Old asset's book value	$2,000	
Trade-in allowance		$(2,300)
List price of new asset		6,000
Cash payment		$3,700

The new asset would be recorded at $5,700 ($3,700 plus $2,000 old asset's book value) or the book value plus the additional cash payment in acquiring the new asset. For tax purposes, no recognition is given to gain or losses resulting from trading one asset for another until final disposition is made.

Other assets include intangible items such as patents, franchises, organizational expenses, goodwill, miscellaneous funds for special purposes, and advances. If these assets have a limited life, the cost should be allocated over the useful life. A patent gives the company a right over a period of time up to a maximum of 17 years, but its useful life may be only 10 years. Therefore, the cost would be allocated over 10 years.

Current liabilities are obligations that are to be paid within a short period or at least during the coming year. These occur from purchasing inventories, services, or supplies, which are the accounts payable. Other current liabilities are notes payable, taxes payable resulting from withholding income tax, and payroll taxes.

Long-term liabilities are notes, bonds, and mortgages that will not be paid within a year. There are other liabilities such as deferred revenues which result from advance payments.

Owners' equity includes the amount invested by the owners and the retained earnings that have not been distributed to the owners in the form of dividends. Net losses in excess of retained earnings are deductions in owners' equity and are classed as a deficit. The owners' equity is listed separately because it shows the amount by which the assets can shrink before the creditors' position is jeopardized.

Types of Budgets

The cash budget is a useful accounting tool for the manager. Capital may be invested in either equipment (fixed capital) or raw materials, finished goods, and cash. Working capital is the difference found by subtracting current liabilities from current assets.

Management of working capital is one of the major concerns of financial executives. The availability of cash is a constant problem in achieving the profit goals of the enterprise. If working capital can be kept to a minimum, more money will be available for investment and income-producing assets. Therefore, proper management of working capital involves an optimization of the amount required to meet the current operating needs.

Working capital is available as cash, receivables, and inventories. Cash is liquid and available any time. Inventories cannot produce cash until they are sold and receivables are collected. If they are sold, this produces delivery and supplier problems within the corporation and therefore does not necessarily represent a good way to secure operating cash.

Probably more business failures, particularly in small businesses, occur as a result of a poor cash position than for any other single cause. As it should be, this is attributed to poor management. Therefore, the cash budget assumes great importance as a management tool. With it the manager can provide a smooth flow of funds so that the efficiency of the business can be improved. If sufficient funds cannot be generated internally, the financial executive can plan to secure them from other sources. If, however, a surplus of cash is anticipated, he can prepare to invest it for maximum return.

The cash budget may be used either as a simple forecasting device or as a means for effective planning. As a forecasting tool, it allows the financial manager to match cash inflows and outflows and provide for any deficiencies. When used for planning, it assists in estimating the mix of inflows and outflows that will provide the greatest contribution to the profitability of the enterprise.

The revenue and expense budget spells out plans for revenues and operating expenses. The most basic part of this budget is the sales forecast. The sales forecast is the cornerstone of planning for the business enterprise. By virtue of this, it is also the foundation of budgetary control. From the sale of products must come the principal income to support the operating expenses and to provide the profits.

The operating expense budget is broken down into a series of classifications that vary from one organization to another. It includes such things as direct labor, material, supervision, rent, heat, power, travel, and entertainment. The various detail line items can become the control feature for the manager using this type of budget.

Some budgets are based on the balance sheet and are referred to as *capital budgets*. These budgets forecast the status of assets, liabilities, and capital accounts at some particular time in the future.

The capital budget outlines planned capital expenditures for plant machinery and equipment. These budgets are important because they involve significant funds and will affect the operation of the corporation over a long period of time. Since capital resources for investment are a major factor in the growth of the corporation, these decisions must be tied in with long-range corporate goals.

One type of budget that is useful is that devoted to *material and products*. These budgets may be based on measurable units such as direct labor, hours measurable per unit of product, machine–hour cost, units of raw material per pound of finished product, and similar ratios. These comparisons are very useful to the superintendent and plant manager in controlling their operations.

The budgets discussed above may be combined together for ease of review into a master operating budget. A portion of the information provides an income statement, and the results of this plus additional information provide the balance sheet data. This information provides the manager with a monthly guide to review the efficiency of the operation.

At this point a few *words of caution* must be offered. The primary purpose of budgets is for planning and control. If they are too detailed and are enforced too rigorously, they may actually have a negative influence on the growth of sales and profits. They should be used with a modicum of common sense. They are a guide, and if the internal and external environmental conditions change, they should also be subject to change. Management, however, sometimes gets the feeling that the budget is totally inflexible and that any deviation from it is wrong. This inflexibility may stifle the initiative of the staff and also perhaps cause the missing of excellent opportunities for new products or increased sales.

The manager will probably encounter the *flexible budget*. Most budgets today are being designed to vary with the sales volume. Certain items on the budget are designated as variable and are influenced by the output volume. Others are designated as fixed such as depreciation, taxes, and insurance and are independent of volume. The flexible budget can be readily varied if indication of greater sales is in evidence. This avoids failing to move to meet increasing sales because to do so might cause certain items of the budget to exceed anticipated values. It also leads to a more common-sense approach to the entire budgeting procedure.

Another type of budgeting that has gotten considerable attention in recent years is *zero-based budgeting*.[3] In normal budgeting the tendency is to look only at the differences from previous periods, which results in a continual upward trend of costs. The idea of zero-based budgeting is to start from ground zero at the beginning of each period and build up the needed resources to meet the goals for that period. It has been most effective in staff support areas such as personnel, research and development, marketing, and finance. The programs of these areas are reviewed in relation to their contribution to the enterprise. The objective of this approach is to force the manager to plan his program completely for each new period. It forces him to consider and evaluate each element of his program against cost-benefit criteria.

Procedures for Preparing a Budget

The first step in preparing a budget is the sales forecast, which is normally made by product line. If the engineer manager is involved in preparing a budget for a plant, project, or product line, the sales forecast will determine the volume of production required from the operation to provide for the sales requirements. Some of the techniques used in sales forecasting were discussed in Chapter 3.

The next step is preparing the production budget based on the sales forecast. The variations of the sales forecast within the accounting period may require increased inventories during certain months if the plant is to operate at a fairly uniform production rate during the period. Preparing the production budget involves determining the level of expenses during the period. Two types of expenses must be determined: variable and fixed. Fixed expenses will not change regardless of production. They include insurance, taxes, and support staff. The variable expenses will vary with production volume. The two major variable expenses are direct labor and materials. Other variable expenses may include utilities (or a portion of them), transportation, and supplies. Sales and production must be coordinated, for neither function can be planned separately.

The third step involves developing a *capital budget,* which includes capital resources required to maintain present production capacity, to make production more efficient, or to expand capacity. The capital budget includes the total cost

[3]P. A. Pyhrr, "Zero-Base Budgeting," *Harvard Business Review* (November–December 1970), 111–21.

of the installed facility including engineering design, delivered equipment, and installation cost. Because these costs will be recovered over a period of years, they are depreciated as discussed earlier.

Budgets should use the same accounting format as the accounting department uses at the end of each month, quarter, or year, so that comparison can be readily made. Sometimes a rolling budget is used. For example, if a year's budget includes 12 monthly budgets, then when the present budget month is over, a budget for the same month next year is prepared. There is always a budget for a year in advance and the budget is always up to date because the future months' budgets are adjusted as circumstances warrant. Under this method, budgeting is a continual process.

Accrual Concept. In most firms the accrual method is used in recording revenues and expenses. This method recognizes expenses when incurred by the firm, not necessarily when a cash transaction occurs. Similarly, revenue is recorded when it is earned, not when cash is received.

Generally, goods and services are sold on credit terms, payments being received at a later date from the customer. This gives rise to what are known as *accounts receivable.* Management must determine the normal credit terms and how long a customer will be allowed to exceed this time without penalty. When payment is received, this decreases the accounts receivable and shows not as revenue, but as additional cash.

Similarly, expenses are recognized and recorded when they are incurred by the firm. Supplies are purchased on credit terms. At the time of purchase they are recorded as an asset and as a liability to the creditor (accounts payable). When payment is made for the supplies, the liability is reduced. As supplies are used in the operation, the portion used is recorded as an expense and the supplies asset account is reduced by a corresponding amount.

Depreciation. This item deserves special attention by the engineer manager because confusion often exists regarding depreciation and its accounting treatment. The expense due to depletion of asset value charged to the operation of an asset during the accounting period is depreciation. When depreciation expenses are recorded, they reduce the asset's value which is recorded in the balance sheet as follows:

Assets (initial cost)	$85,000
Less accumulated depreciation	$24,450
Assets current value	$60,550

But the nature of such accounting practice is often misunderstood, as discussed earlier. The misunderstanding arises because it is assumed that depreciation entries produce funds for replacement of the depreciable assets. This is a false assumption. Depreciation charges are recorded each accounting period during the asset's life but in no way does this create a cash reserve or even affect the cash account. Nor does this provide a fund for the replacement of the asset, since the procedure does not set aside cash to replace the asset.

Analysis of Accounting Information

There are many control techniques that are used by the engineer manager and that complement the accounting and budget controls. Because there is extensive information on these procedures in accounting and nonaccounting books, only two will be discussed here.

1. *Break-even analysis.* A very basic control technique is the break-even analysis. This analysis shows the relationship of sales volume to expenses and it indicates the volume necessary to cover expenses. At smaller volumes there would be a loss, and at greater volumes there would be a profit. The break-even point is the point at which there is neither a profit nor a loss. This is important to the engineer manager in determining the production level at which he must operate to avoid a loss. Break-even analysis emphasizes (1) the effects of additional sales or costs and in turn profits, (2) impact of fixed costs, and (3) the marginal concept—the effect of incremental changes on volume.

The break-even chart (Figure 10–2) represents these relationships as straight lines. It is well known that the relationships are not linear, but for any given operation it can reasonably be assumed that the chart will operate within a given range where volume and expenses are usually of a linear nature. Therefore, sales and expenses can be represented by straight lines. The engineer manager's objective should be to determine the volume required to obtain a profit or to stop losses for a given plant or operation. This is basic information for his use in the decision-making process. In addition, it shows how profits are affected by changes in selling prices, variable expenses, and fixed expenses. What may appear to be a very small per-unit saving may be the difference between a profit or loss for a production facility during the year.

When there are high fixed expenses for relatively low-priced items, profits can be increased by increasing production. But reduction of fixed expense will reduce the break-even point and increase the profit potential.

Break-even analysis allows the engineer manager to compare his operation with past accounting periods and with other similar operations to determine if variable and fixed expenses need to be examined. For the engineer manager, there are very few operations in which variable and fixed expenses cannot be reduced by implementing more efficient engineering management practices. The break-even analysis indicates areas toward which the engineer manager should direct his attention and efforts.

2. *Return on Investment (ROI).* This control technique measures the success of a production unit by the ratio resulting from the following equation:

$$\text{ROI} = \frac{\text{Earnings}}{\text{Sales}} \times \frac{\text{Sales}}{\text{Total Investment}}$$

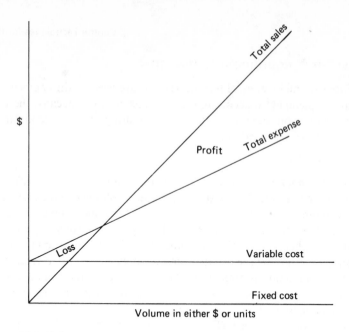

FIG. 10–2. Break-Even Chart

This equation considers the interlocking relationship among (1) earnings on the sales, (2) sales volume, (3) investment on assets, and (4) ratio of sales volume to assets. These relationships are expressed as follows:

$$\text{ROI} = \text{ratio of return on sales} \times \text{asset turnover}$$

ROI [sometimes referred to as *return on assets* (ROA)] was made popular by the DuPont Company which has used the technique for many years. Many companies have now adopted the technique. ROI is a measurement of what the operation can earn on the investment allocated. The objective is to determine return on investment employed. This is a measure of the attractiveness of the investment to the investor. The fundamental assumption of the technique is that investment is a critical resource in all operations and the lack of adequate investment limits corporate growth. Some of the more common applications of ROI techniques are:

A. Measuring the profitability of the operation or the organization
B. Providing information for planning and controlling operations
C. Evaluating and comparing the firm's assets, earnings, and sales relationships with those of other firms
D. Assisting in developing alternative uses for investment monies
E. Determining areas that need attention

To calculate ROI for various segments of the operation, these segments must be treated as profit centers by the accounting department. If the ROI is low for a given operation, an investigation can be made to determine the cause. The cause may be that the assets are not being used efficiently. Perhaps the spread between selling prices and expenses needs to be widened by increasing prices or reducing expenses. The engineer manager must realize that he not only has responsibility for operation but also responsibility for the use of the assets.

The engineer manager can estimate ROI for various product lines or operations that are not profit centers by the allocation of sales expenses and investments. In many organizations, expenses are maintained by product and operation. Cash, accounts receivable, and administrative expenses may be allocated by sales or production volume. A more difficult task is allocating asset usage to the product line.

Before PERT came into existence, the most extensively used scheduling and control device was the Gantt chart.[4] This chart was developed by one of the pioneers in the field of management, Henry L. Gantt. By graphing operations on a sequential time scale, it is easy to understand which operations must be completed before or simultaneously with other operations.

PROGRAM EVALUATION AND REVIEW TECHNIQUE (PERT)

Three forms of Gantt charts can be used: (1) progress charts, which give a sequential schedule, (2) load charts, which show operations assigned to workers or machines, and (3) record charts, which record the actual times involved and any delays that may be incurred. The production plan shown in Figure 10–3 illustrates a simplified progress chart. The operations to be planned are listed along the vertical axis. The horizontal axis shows the time necessary to complete each operation.

	Weeks													
	1	2	3	4	5	6	7	8	9	10	11	12	13	14
Scheduling	▨													
Engineering	▨	▨												
Order			▨											
Receive materials				▨	▨									
Assemble						▨	▨	▨	▨	▨				
Inspection											▨			

FIG. 10–3.   Gantt Progress Chart of Production Plan in a Mechanical Device

The Gantt chart graphically represents what is planned and it allows progress to be recorded. It reveals a comparison between planned and actual operations.

PERT is both a control and planning technique that can have many applications in scheduling construction and engineering project activities. PERT was developed as a technique for aiding the Special Projects Office of the Department of the Navy in 1958.[5] It is a systematic method for planning, organizing, and evaluating all resources so that they are available on a timely basis to meet all

---

[4]L. P. Alford, *Henry Laurence Gantt: Leader in Industry* (New York: American Society of Mechanical Engineers, 1934), pp. 207–23.

[5]*Summary of Minutes,* Meeting of Contractors PERT Reporting Personnel, U.S. Naval Weapons Laboratory, November 16, 1960, p. 5.

exigencies of the project. In a practical sense, PERT is budgeting, for it considers time events, but it is also a network analysis (Figure 10–4). Each circle represents an event, a subsidiary plan whose completion can be measured at a given time. Examples of events are the point where funds are released, a contract is awarded, or specifications are released. Each arrow represents an activity, the time-consuming elements of a program. Examples are design, analysis, and fabrication.

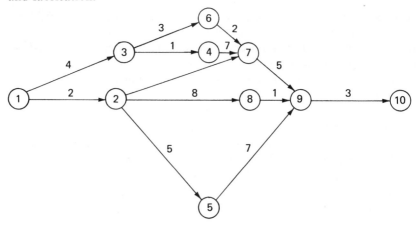

FIG. 10–4.   PERT Network Diagram with Time Estimates

The activity time is determined by three time estimates: "optimistic" time, the "most likely" time, and the "pessimistic" time. These estimates are developed because it is difficult to determine time accurately. These estimates are averaged, with special weights given to the most likely estimate and a single estimate time used:

$$t_e = \frac{\text{optimistic time} + 4 \text{ most likely time} + \text{pessimistic time}}{6}$$

Where t_e = average estimate time. After the estimated times are calculated and recorded on the arrows, the critical path is computed; this is the sequence of events that takes the longest time, and therefore has the least slack time. Through this path the total activity time can be determined for the project. In Figure 10–4 the critical path is 1–3–4–7–9–10.

In practice, the critical path changes as events are delayed in other parts of the project. Other paths may be nearly as long and should be recorded and ranked in order. It is possible one of them might become critical during the course of the project. All events can be watched to assure that they receive proper attention and that the project remains on schedule. Some projects have a large number of events, perhaps hundreds of thousands. For these, PERT computer programs are available. For very large projects, subordinate networks are prepared and a summary can be made for top management.

The basic steps involved in constructing the network are as follows:

1. Determine the objective and then determine the events that must occur in order for this objective to be reached. For example, the objective might be to design, purchase, and install a new distillation column to perform specific functions in a certain plant. Some of the events involved in this operation would include release of specifications for bid, awarding of a contract, preparation of a site, receiving equipment, installation of equipment, and testing and acceptance of the facility by the customer.

2. Starting at the left side, place circles in their proper relationships to represent events. Join these circles by arrows. Some of the events can take place while other events are in progress; these will be noted by event location showing concurrent actions.

3. For each activity indicated by an arrow, estimate the expected time to accomplish that activity. Record this figure on the network.

4. Starting at the left with the starting event, calculate and record the expected completion time for each event. This can be determined by the expected time for activities as previously recorded.

5. Now starting at the extreme right, work backward through the network, calculating the latest allowable date for each event and record. This means the latest allowable time that a certain event can take place without affecting the desired project completion date. Thus, this allowable date is both the latest allowable completion date for all preceding activities as well as the latest allowable starting date for the most critical task that follows it. It is represented by the difference between the estimated time in the activity and the succeeding event's latest allowable date.

6. Next calculate and record the slack time for each event in the network. This represents the difference between the total expected time for the event and the latest allowable time for that event. It represents the extra time available for noncritical events.

7. Now determine the critical path through the network. This will be the longest path available and it is found by connecting the events that have zero or negative slack. The summation of these critical events controls the ending date of the project.

PERT/Time

PERT/time is a control method that is widely used by both large and small industries.

It has the following *advantages:* It is extremely useful in the construction industry for planning, coordinating, and control of all types of projects. It has also been used successfully in R&D projects. PERT/Time owes its popularity to the discipline of planning. It forces a manager to plan because it is impossible to make a time-event analysis without planning and seeing how the pieces fit together. Further, the manager must think of all the events necessary in order to achieve a specific goal. This helps him to avoid missing critical items. It also tends to integrate the entire organization because it forces various subordinate managers to plan the event for which they are responsible.

PERT/Time also has the advantage that it concentrates attention on the critical elements that will need correction. It forces the manager into a careful analysis of his overall project. It actually provides a forward-looking control system in that it becomes evident that delays on one event will affect the entire project.

The *disadvantages* of PERT/Time include the following: The usefulness of PERT/Time is limited when dealing with nebulous programs. If the network is to be of value, one has to be able to make reasonable guesses on how much time the activities will consume. It is also useful primarily for a nonrecurring set of events in which a series of individual events combine together to provide a final event.

Preparation of the PERT/Time network can be a tedious and demanding task if done manually. It means that a project of any complexity must use a computer. Fortunately, software has been developed to handle projects involving several thousand activities. Nevertheless, the manager still must do the rigorous thinking and planning that develops the events necessary to be scheduled. Finally, PERT/Time does not speak to an extremely important factor—cost.

PERT/Cost

Networks have been developed using the basic concept of PERT in which costs of activities have been introduced. This adds greatly to the complexity of the situation. If there are many activities involved, it is difficult and time-consuming to make cost estimates of all of them, and further data are not available for comparison for control purposes unless job accounts have been established for each activity. One way out of this difficulty is to group activities into work packages and accumulate costs in this manner. There is also the question of what to do about overhead costs that are not directly related to the activities. Further, there may be a lack of balance between a PERT/Time network and a PERT/Cost network since they may progress forward at different rates. Nevertheless, it does provide an effective managerial tool for planning and control and, while PERT in any format is not a cure-all, it certainly does force the manager into that important area of planning.

With both PERT/Time and PERT/Cost, it is possible to trade time units for costs, or vice versa, thereby obtaining the least cost possible. This relationship is illustrated in Figure 10–5. As costs increase to a certain point, time decreases, and vice versa. At times, it may be best to increase cost for a critical activity, and thereby reduce time.

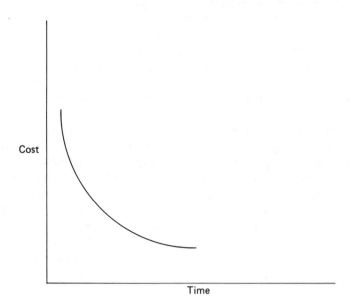

FIG. 10-5. PERT/Time and PERT/Cost Relationship

The PERT technique is one of the most successful engineering management aids, especially in planning and controlling lengthy and costly projects. Scheduling of these projects is difficult without a systematic approach such as PERT because of the large number of relationships that must be considered.

Once plans have been implemented, the engineer manager is responsible for using control techniques to measure progress toward planned objectives. The control process includes establishing standards, measuring performance against these standards, and correcting deviations from standards and plans.

SUMMARY

Feedback is the important ingredient of the control process. To use this principle, the process must be appropriate to what is being measured, be compatible with abilities of people, and be realistic.

Forward-looking or feedforward controls are gaining in importance.

An effective control system must be timely, economical, understandable, flexible, and individually tailored and provide corrective action. Accounting systems provide major forms of control. The income statement is fundamental to the process since by collecting sales and expense information it can determine if the operation is either making or losing money. The data can be broken down in a number of ways, allowing one to examine profit or loss by product line if needed. It also provides a framework for comparison with competitive firms in the industry.

The balance sheet accumulates accounting information in a manner that the financial health of the organization can be reviewed. It lists the firm's resources and the creditors and owners of these resources.

There are a number of budgets useful in control, notably, the cash budget, revenue and expense budget, and capital expenditure budget. To prepare the budget requires the help of many departments throughout the company. The capital budget is exceedingly important because it controls the future of the company.

Break-even analysis is useful to the engineer manager in determining the critical production volume to assure break-even on a project. Return on investment (ROI) is important, especially in these days of high interest rates.

Program Evaluation and Review Technique (PERT) is a technique of network analysis that is useful in scheduling and controlling projects, especially large, complex projects. These latter will probably need the aid of a computer.

PERT/Time has the advantage of forcing the engineer manager to think through the activities needed to achieve a final event. PERT/Cost has been developed to take care of those special cases in which cost is particularly important.

Important Terms

Balance Sheet:

Accounting information for a specific point in time that shows the firm's resources and the creditors and owners of the firm.

Control Techniques:

Techniques used to measure progress toward planned objectives, to identify deviations from plans, and to take corrective action.

Corrective Action:

Action taken to adjust the system so that output can match the goal.

Feedback:

Information on whether or not the output of the system is meeting the established goal.

Income Statement:

Accounting information for a period of time that is assembled to measure profit or loss.

Standards:

Verifiable goals or objectives against which progress can be measured.

For Discussion

1. What are the three steps that make up the control process?
2. What is the importance of setting standards?
3. Why is an effective feedback system important to management control?
4. Name five characteristics of a good control process.
5. Discuss the importance of forward-looking control.
6. Discuss the meaning and purpose of the income statement as an accounting control technique.
7. Discuss the elements that make up a balance sheet.
8. Discuss the construction and use of the cash budget.
9. What constitutes the capital budget and why is it important?
10. Discuss zero-based budgeting.
11. Outline the steps required for the preparation of a budget.
12. Describe the construction and use of a break-even chart.
13. How do you calculate return on investment and of what value is it in the control process?
14. Describe the elements that make up a PERT network.
15. What are the advantages and disadvantages of PERT/Time?

Case 10-1

Who Controls the Cost?

Rex Smith had a reputation with top management as a man who was willing to cooperate. If the management had a new idea, they would say, "If Rex can't make it work, nobody can."

The company had never examined the units closely for costs. But recently a new comptroller was hired and every operation including R&D was carefully looked at by his people. They concluded that Rex had gotten sloppy about the use of overtime, control of project costs, use of supplies, and excessive waste in the laboratories. Consequently, Rex's operation was issued a budget for these at the beginning of the following month.

At the end of the month Tom Johnson called Rex into the office. "Rex, in checking over your budget for R&D for the past month, I find you are over your budget by 14.2%. Will you see if you can bring this back into line next month?"

The following day Rex called together his staff. "Staff, we have got a real problem to cope with. We have an order from the vice president to cut costs 15%. We will have to stop overtime for awhile and watch operating costs carefully. This requires us to especially watch the use of our supplies." Staff members agreed they would all do their part and told Rex to make the necessary reductions and tell them what to do.

At the end of the next month, to his surprise, Rex was again called into Tom Johnson's office. "Rex, I hate to bring up the matter of cost overrun again. I know that you have a good operation and are making a real contribution to our profit, but your operation made no progress in reducing costs last month. It is absolutely necessary that you reduce costs this month."

Rex called the staff together the same day. "We must reduce these costs because we are looking bad in the eyes of Tom Johnson. I am depending on each of you to help reduce costs to within budget."

At the end of the month Rex received actual operating costs compared with budgeted costs. There was an overall reduction of 2.5%. Nothing was heard from the vice president's office. But by the end of the following month, costs were back where they were when the budget was first issued. Rex was surprised when Tom called him in and asked why he had been so unsuccessful. Rex replied, "I can't understand it. They all agreed to work extra hard to reduce costs."

1. What did Rex do wrong? What did he do right?
2. What is his relationship to his staff? Do they help him?
3. What must Rex do to get costs within the budget?

Case 10–2

Employee Work Control

Hernandez Garcia, head of the XL-19 department of Metal Products Company, was having some problems. His department had some moderately complex assembly operations using semiskilled workers at relatively low-wage rates. He felt that during the past year there had been a gradual decline in employee morale and in employee productivity. Taking what he felt was a reasonable approach to the problem, Garcia asked his supervisors to keep a closer eye on the activities of the workers. They were to make sure that the workers applied themselves while on the job and that they did not extend morning and afternoon breaks or prepare to leave the job early. After a month or two of this Garcia failed to see any significant change in productivity.

Feeling that perhaps something was wrong with the production lines, he asked Rex Smith to send over an engineer to do some efficiency studies on his assembly lines. One thing was soon evident to the engineer. Because of the sequential actions of some of the employees in the assembly operation, the line tended to run only as fast as the least productive employee.

The engineering management analyst also noted something else. Mr. Garcia had had to authorize considerable overtime in order to meet delivery schedules as a result of the lowered productivity. The analysis showed that the workers tended to work slower at the beginning of the month so that they could be sure that overtime would be required by the end of the month. As a matter of fact, even some of the most senior and better-skilled employees were doing the same thing.

Garcia questioned what he should do. He was somewhat reluctant to fire the employees; in addition, he found it difficult to actually catch an employee slowing down the line. He was afraid that if he were to fire an employee without having adequate information, he would have to take the employee back and this might have an even worse effect on morale and productivity.

1. Describe in detail the control dilemma that exists.
2. Are Garcia and the workers getting the same feedback?
3. What do you recommend that Garcia do?

For Further Reading

ARGYRIS, CHRIS, "Human Problems with Budgets," *Harvard Business Review* (January–February 1953), 97–110.

BONCHANSKY, JOSEPH P., "Cost Control for Program Managers," *NAA Management Accounting* (May 1967), 16–24.

BOULTON, W. R., "The Changing Requirements for Managing Corporate Information Systems," *MSU Business Topics* (Summer 1978), 4–12.

BUFFA, ELWOOD S., *Modern Production Organization Management*. New York: John Wiley & Sons, Inc., 1980.

CANNON, C., and D. A. NADLER, "Fit Control Systems to Your Managerial Style," *Harvard Business Review* (January–February 1976), 65–72.

COOK, T. M., and R. A. RUSSELL, *Production/Operations Management*. Englewood Cliffs, N.J.: Prentice-Hall, Inc., 1980.

GRANT, E. L. and W. G. IRESON, *Principles of Engineering Economy*, 5th ed. New York: The Ronald Press Company, 1970.

KOONTZ, HAROLD, and R. W. BRADSPIES, "Managing Through Feedforward Control," *Business Horizons* (June 1972), 25–36.

MCFARLAND, W. B., *Manpower Cost and Performance Measurement*. New York: National Association of Accountants, 1977.

MOCKLER, R. J., *The Management Control Process*. New York: Appleton-Century-Crofts, 1972.

MYERS, LARRY R., "Effective Engineering Budgets," *Machine Design* (December 8, 1966), 152–55.

NEWMAN, W. H., *Constructive Control: Design and Use of Control Systems*. Englewood Cliffs, N.J.: Prentice-Hall, Inc., 1975.

RASMUSSEN, SVEIN G., "Budgeting and Control of Development Projects," *Machine Design* (August 29, 1968), 78–81.

SCHMIEG, HARRY J., "Control of Overhead with a Variable Budget in a Research Operation," *NAA Bulletin* (August 1959), 45–56.

SEILER, ROBERT E., *Improving the Effectiveness of Research and Development*. New York: McGraw-Hill Book Company, 1965.

TOSI, HENRY L., "The Human Effects of Budgeting Systems in Management," *MSU Business Topics* (Autumn 1974), 53–63.

VANCIL, R. F., "What Kind of Management Control Do You Need?" *Harvard Business Review* (March–April 1973), 75–86.

WHITEHOUSE, G. E., *Systems Analysis and Design Using Network Techniques*. Englewood Cliffs, N.J.: Prentice-Hall, Inc., 1973.

EFFECTIVE
COMMUNICATION

One of the major tasks of the engineer manager in performing his managerial role is to transfer information to others. He must acquire new techniques to communicate both verbally and in writing with groups of people at all levels in the organization.

CHAPTER
HEADINGS
- ☐ Understanding the Information Process
- ☐ Communication System Concepts
- ☐ Types of Systems
- ☐ Developing an Information System
- ☐ Designing the Information System
- ☐ Developing an Appropriate Style
- ☐ Developing Effective Skills
- ☐ Effective Listening
- ☐ Effective Communication
- ☐ Summary

LEARNING
OBJECTIVES
☐ Learn how to develop and use the communication process.
☐ Develop knowledge of the various information systems.
☐ Gain insight into how to develop an information system.
☐ Learn appropriate styles of communication.
☐ Learn the importance of effective listening.
☐ Develop understanding of how to improve oral and written communication skills.

EXECUTIVE COMMENT

GORDON H. MILLAR
Vice President
Engineering
Deere & Company

Gordon H. Millar joined Deere & Company in 1963 as director of research. He organized and directed the Technical Center until 1969 when he was appointed assistant general manager of the Waterloo Tractor Works in charge of design and development of all John Deere engines. After an assignment in Switzerland, he was named in 1972 to his present position as vice president, Engineering.

Effective Information Systems

The creative output of engineering is only of value when it is ultimately communicated and implemented. Many times this is one of the toughest challenges for engineers, even though their disciplines allow for more exacting communications. These communications take two basic forms. We could call one form interpersonal communication, which involves communicating general rationale and objectives, managerial reason, group problem-solving activities, and simple personal dialog. The other form of communication involves the more precise area of data transfer, the communicating of routine information from source to ultimate user. It is in this area that information systems play a key role. Through reliable formulation, engineering time is freed for more creative challenges and efforts.

In developing information systems we can apply the same degree of technology as we do to other engineering problems. Unfortunately, the tendency is to accept current communication techniques without question. This often allows the communication of engineering results to become a bottleneck that jeopardizes the efficiency of implementation. The computer is the primary new tool in modern information systems. But in applying the computer to communication, we can suffer unintentionally. The power of the computer tends to be focused on complexities, and simplification as an objective and a potential result is frequently lost.

A successful information system is easy to use. If it is complex or cumbersome, regardless of its power, potential users will seek other alternatives. Information systems must also be designed to assume that there will be exceptions to the rule, rather than to optimistically hope they will not arise. Information systems must also always be expandable to accommodate new needs.

Whatever the situation, we can never forget in designing the information systems that we are dealing with people. The personal traits of humans are in fact more important in the design of a complex computerized communication system than they are in one-on-one personal systems in which recourse to explanation and iteration is always possible.

One of the major tasks of the engineer manager in performing his managerial role is the ability to transfer information to others. As an engineer, he used scientific symbols and equations to transfer information, but as an engineer manager he must acquire additional techniques to exchange information with new groups of people at all levels in the organization.

For the engineer manager, developing an effective method of communication is a most important problem. As an engineer he used blueprints, equations, drawings, and symbols to transfer information in his daily work; but as a manager he must add additional techniques, mainly oral and written communication processes. To realize the importance of communication, it has been estimated that in his new role he will spend approximately 90% of his time communicating in one form or another.[1]

The purpose of communication is to transfer an image or idea from one person to another, which is then recast into his own working idea.

A message may take the form of words, bodily gestures, pictures, diagrams, or writing, and it may be encoded and transmitted through a channel. The receiver must then decode the message in terms that have meaning to him and that he understands. The process is not complete until feedback is given to the sender (Figure 11–1).

UNDERSTANDING THE INFORMATION PROCESS

There is reasonable evidence that if an organization is effective in its communications, it will be effective in its overall operation. Chester Barnard, one of

[1]Fred Luthans, *Organizational Behavior* (New York: McGraw-Hill 1973), p. 234.

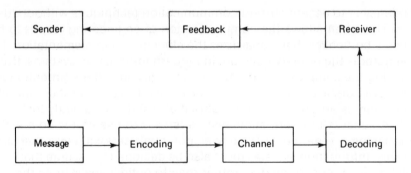

FIG. 11-1. The Communication Process

the early management theorists, has indicated that communication is the key to management.[2] He states that the coordination of efforts essential to a system of cooperation requires an organized system of communication. He indicates that the function of executives is to serve as channels of communication. Therefore, it is important for managers to understand how to communicate and what tools and media to use. In understanding the communication process, the engineer manager must consider such basic factors as the following:

1. The *background* of the people receiving the message, which includes family, ethnic groups, religion, education, nationality, and even region of the country. An individual receiving a message interprets it within the framework of such cultural factors, whether he is conscious of it or not. But when the engineer manager has a different background, there may be difficulty in communicating. The more he has in common with another individual, the greater the likelihood that he will be able to communicate effectively.

2. *Group relationships.* People tend to identify with certain groups and develop loyalties and personal relationships. These values and attitudes of the group tend to influence the way an individual perceives messages from the engineer manager. Therefore, messages are interpreted in the light of the manager's relationships with the group.

3. *Expectations.* When the manager sends a message, the receiver always has certain expectations; in fact, he interprets the message according to the expectation that he has perceived, perhaps gained from his prior experience.

4. *Education.* The level of education makes a difference in the receiver's interpretation of messages. Other factors affecting his interpretation include the individual's expertise, his use of technical data, his method of analysis, and his degree of sophistication.

5. The *situation,* which influences the way a message will be interpreted. If the individual is always criticized when called in to the manager's office, then this is

[2]Chester Barnard, *Functions of the Executive* (Cambridge, Mass.: Harvard University Press, 1938).

what he will perceive, even if he is praised. The same is true with meetings. If they are perceived as a waste of time, then they are interpreted in that light, even if something productive does occur.

6. *Standards* of individual ethics and ideals, which strongly influence how the message is interpreted. Varying viewpoints on honesty and thrift cause individuals to interpret messages in different ways.

7. *Personal welfare.* When the engineer manager's message directly pertains to the personal welfare and interest of the receiver, greater attention is given to the message.

When receiving messages, one tends to interpret them based on the above factors; therefore, distortions of the engineer manager's messages often occur. But by being aware of these probable distortions, he can more effectively reduce them.

In developing an effective communication system, the engineer manager must first determine what he "needs to know." Then he can seriously consider relevant information. The engineer manager must realize that no two people will request exactly the same information. Certainly, there will be common infor-

COMMUNICATION
SYSTEM
CONCEPTS

mation, but each person will demand and require unique aspects of it. Therefore, the communication system must be one the engineer manager can tailor to his individual requirements. This is not to say that there cannot be a uniform communication system for the whole organization, because general guidelines can be set forth whereby each manager can develop his own system. Such systems should be:

1. *Specific.* All information should deal with a single, specific subject. People, especially engineers, require exact measurements, not vague and general statements. People expect the engineer manager to be specific in his communication and he usually insists on receiving specific information.

2. *Intelligible.* If the information is to have value, it should be expressed in language and transmitted in a method that can be comprehended. Most engineer managers are surprised to learn how poorly even the most carefully worked-out information is comprehended. It is his responsibility to express his messages in an understandable way and in language familiar to the people receiving it, i.e., subordinates, peers, and superiors.

3. *Organized.* The information transmission should follow organizational lines of authority and responsibility. Information should not bypass positions in order to contact the ultimate people directly. Exceptions may occur when it is essential to communicate with everyone simultaneously.

4. *Timely.* The speed and timing of communication may be most important. Efforts to obtain intelligible and specific communication may have disadvantageous effects on the organization.

5. *Informal.* To achieve an effective communication system, the engineer manager utilizes the informal organization to complement the formal organization. At times, the formal channel is inadequate and unreliable for messages. Often the engineer manager will authorize a subordinate to establish contact outside the formal organization in order to assure the transmission of information.

6. *Appropriate.* Information should pertain to the interests and needs of the individual. The engineer manager needs to provide information that is warranted, such as the facts of the situation, both pleasant and unpleasant. He should always ensure that his employees are fully informed on policies, plans, and nonmanagement information.

TYPES *OF SYSTEMS* Many communication systems developed for an organization rely principally on nonhuman systems. Therefore, the engineer manager should be familiar with both human and nonhuman communication systems. In most cases, the engineer manager considers only written and oral communication, but he should take advantage of all types of information transmission systems available. The most common information systems are described below:

1. *Electronic Data System.* One of the most rapidly growing information systems that the engineer manager can use effectively is computer processing. This equipment has made it possible to communicate immense quantities of data on a regular schedule and effectively handle information that was impossible before. In fact, many information systems rely principally on computers. These include accounting, production, control, inventory, scheduling, forecasting, R&D, to name a few which are expanding at a rapid rate in all areas of the organization. The advantages of electronic data systems are:

A. *Speed.* Information can be provided in a fraction of the time it would take to produce it with conventional methods.
B. *Storage.* Vast amounts of information can be stored in relatively small areas and can be retrieved easily.
C. *Accuracy.* Information can be handled with minimum errors.
D. *Efficiency.* Information can be handled repeatedly and transmitted very efficiently.
E. *Data Reduction.* This is indispensable for reducing large masses of data quickly and producing new information from it.

These make it possible for the engineer manager to make decisions based on considerably greater information than is possible without electronic data processing.

Electronic data systems do have disadvantages that the engineer manager must recognize and understand. These can be summarized as follows:

A. High equipment and programming costs can make reports expensive.
B. Demand must be sufficient to justify the purchase of a system.
C. Information acquires status as computer printout, even though it may be useless or incorrect.
D. Information is only as good as the human who programmed and entered the data; the information must be translated and designed to meet the needs of the engineer manager.
E. Poor and inadequate planning and lack of definite uses result in undesirable end products.

The engineer manager can overcome all of these disadvantages by proper planning and use of the system.

2. *Formal Information System.* In any organization there are formal chains of command through which official information must flow. Most written communication follows these formal channels and reflects the superior–subordinate relationship. The engineer manager must realize that the formal channels frequently become bureaucratic and slow down information transmission. Formal information is communicated by such methods as:

A. Written procedures and policies
B. Letters and memos
C. Company newspapers and magazines
D. Company handbooks and policy manuals
E. Information booklets and training manuals
F. Annual reports
G. Filmstrips and slides
H. Accounting statements
I. Special messages
J. Legal documents
K. Job descriptions

3. *Verbal Communication.* Verbal communication has long been a tool essential for the manager to carry out his work. Certainly, verbal communication permits feedback, two-way communication, stimulates interest, and fosters cooperative relationships. The manager must master the ability to use verbal communication effectively.

One special area of oral communication for the engineer manager involves the news media. If he is a plant manager or district sales manager, he is the company representative. If he is project manager for the construction of a new plant, he may find himself the focal point of the news media. When he makes statements to the media, he should:

A. Be aware of company policy in dealing with the media, and particularly for employees in his position.

B. Try to have a prepared statement covering the event that has made news.

C. Make sure any questions are understood.

D. Think as he speaks.

E. Not answer more than is asked. Be brief.

F. Be open and honest.

G. Have available supporting data.

H. Keep superiors advised of media contacts.

4. *Technical Information System.* Blueprint drawings, technical report models, and pilot projects are all important parts of the engineer manager's information system. These all fulfill a need in conveying information that is difficult to handle by spoken or written words alone.

5. *Informal Communication System.* Every engineer manager has an informal communication system that serves several purposes and needs in his functional area. In fact, an organization would grind to a halt if it were not for the informal communication system that develops with time. It is a very dynamic system and it provides rapid communication in all directions, not just up and down. This system provides a flow of information directly from those who accumulate or generate it to those who need it, regardless of formal organization lines.

The informal communication system is often referred to by the engineer manager as the *grapevine*. Generally, it is a verbal channel. The system is generally structureless, but it goes into operation when members of the formal organization who know each other well enough begin to pass on information dealing with the enterprise. The grapevine thrives on information that is not generally available to the entire group.

Many engineer managers attempt to suppress the grapevine, but suppression only makes it a more powerful system. Because the grapevine is speedy, relatively accurate, and ubiquitous, the manager should consider it as one of his important communication systems. He should learn who the key individuals are in this information system. Then he can give these people the information to be communicated. This is especially important when the grapevine system has been spreading incorrect information.

DEVELOPING AN INFORMATION SYSTEM   The engineer manager must recognize that the environment within the organization also has an effect on the successful communication of information. The effectiveness of the communication system is influenced by clearly defined channels. These channels involve the superior–subordinate relationships and the various information systems throughout the organization. When individuals know with whom they should communicate directly and what information must be transmitted, communication breakdowns can be avoided.

Problems are normally caused by:

1. Difficulties with confused lines of authority
2. Organization structure being altered frequently
3. Organization being established without proper planning
4. Too many levels of organizational authority
5. Wide span of control
6. Interests of engineer manager not directly related to the project
7. Physical distance between individuals utilizing the communication system

The initiative for communication is principally the responsibility of the manager. He must communicate information needed by his subordinates. It is more difficult for subordinates to communicate to the engineer manager because the initiative is up to them. Subordinates must feel a special need and must also feel secure before they will communicate effectively to the manager. Conditions that cause subordinates to be unwilling to communicate with the engineer manager are listed below:

1. They are afraid that their communication will result in negative rather than positive action.
2. Subordinates feel that the engineer manager is unsympathetic and they are distrustful of him.
3. Subordinates feel that they are unimportant and therefore their information is unimportant. The more important subordinates feel, the greater their willingness to communicate.
4. Subordinates do not have a common background with the engineer manager and they do not share his interests.
5. The engineer manager is unavailable.
6. The engineer manager shows little or no desire for information from his subordinates and does not appreciate their participation.
7. Subordinates believe the engineer manager has a negative attitude toward them.

To encourage communication from subordinates, the engineer manager must:

1. Provide an environment that induces two-way communication.
2. Demonstrate an attitude of sincere interest in hearing from his subordinates.
3. Reward his subordinates for their communication efforts.
4. Make subordinates feel their inputs of information are vital, necessary, and appreciated.

5. Request reports and information from subordinates on a regular basis to ensure that adequate communication occurs.

The engineer manager must be concerned with the problems involved in communication with other engineer managers and supervisors at the same organizational level. He must be aware of the needs of his colleagues and should attempt to send and receive accurate and useful information. A spirit of cooperation greatly facilitates the effectiveness of the engineer manager's information system.

Elements of an Effective Information System

Developing an effective information system depends on the engineer manager's taking several positive steps to encourage the improvement of information transfer. Some of the most important steps are listed below:

1. *Clearly Express Messages.* He must examine all of his messages to ensure that he avoids:

    A. Omission of information
    B. Lack of coherence; inconsistency
    C. Poor organization of ideas
    D. Awkward sentence structure
    E. Inadequate vocabulary
    F. Repetition of data and words
    G. Meaningless words and phrases

By watching for these, the engineer manager's information system will have clarity and avoid costly errors and corrections.

2. *Interpret correctly.* The manager receives information from subordinates, peers, and superiors in many formats. He must in turn interpret information from these and provide others with information in a language they can use and understand. In most cases, he cannot pass the same information on to others without restructuring it in an appropriate form so that they can understand and use it. This requires him to:

    A. Have a working knowledge of the receiver's discipline
    B. Have an understanding of the receiver's goals and objectives
    C. Use language and terms the receiver uses, i.e., engineering terms for engineers and accounting terms for accountants.

These skills generally enable the engineer manager and his group to operate at a level of maximum understanding.

3. *Increase retention.* Most people have poor retention; therefore, the engineer manager must reinforce and repeat important items.

4. *Reduce need for reinterpretation.* When successive interpretations (from one person to another) are eliminated, accuracy and completeness are greatly improved.

5. *Use the information system.* The engineer manager must fully examine all information to determine which individual or group should receive it.

6. *Develop trust.* For the engineer manager to gain the trust and confidence of his people, he must issue messages that are consistent, logical, true, and honest. Trust in the engineer manager increases the effectivenes of the information system.

7. *Listen.* The engineer manager must develop the habit of listening to his subordinates and superiors. Many times he fails to listen to the message being sent. Not listening is a common habit with many people who are preoccupied with personal problems, golf scores, or business problems.

8. *Look for unspecified communication.* The engineer manager must always realize that any message he receives will involve implied information. Therefore, he must examine each message for its implication. He must examine each of his own messages from the standpoint of having included sufficient details so that the receiver will not be required to assume or interpret beyond reason. Seemingly straightforward information can lead to a great deal of uncertainty and cause delays or lack of action because the engineer manager has assumed that his subordinate or those receiving the information would understand it.

9. *Keep an open mind.* The engineer manager should not commit himself to a viewpoint before he has received complete transmittal of information.

The engineer manager must undertake specific tasks that are especially important in the designing of the information system. These tasks must be developed as an integral part of the organization. In designing the information system, he should:

DESIGNING THE INFORMATION SYSTEM

1. Design the information system along formal organizational lines.
2. Include the needs of every member of the organization.
3. Design the system so that it is as direct and short as possible.
4. Ensure that the system uses current and valid communication sources.

The objective of the design is to attain and build an integrated system so that the communication process becomes a reality and uses and involves all available personnel, equipment, and other resources of the organization. Certainly the information system's design should complement the organizational structure and operating methods, as well as the role and scope of its goals or objectives. Therefore, the system must attempt to provide an orderly and useful conceptual framework for communication.

Designing the information system can be straightforward, but maintaining effective communications is difficult. The problem is magnified by the spectacular increase in the volume of information available and required by the engineer manager and his group. This includes data and verbalized information on policies, politics, economics, competition, and internal procedures. In the design of the system, the engineer manager must ensure that the information is fully utilized, for in practice, much information is unused by the group. In addition, information frequently does not reach the right people.

Even the best-designed system requires that the engineer manager teach his group how to use the information fully once they receive it. This involves:

1. Analyzing the information being *transmitted by:*
   A. Type
   B. Purpose
   C. Timing
2. Analyzing the "needs" for information by:
   A. Type of information system required
   B. Timing
   C. Purpose
   D. Best sources
3. Developing new information based on 1 and 2 above.
4. Training, including
   A. Meaning of data
   B. Sources of data
   C. Uses of data

This provides an opportunity to eliminate information that is never used. Most people are overwhelmed by unnecessary reports, directives, and memoranda that they do not use or understand how to use. Because each individual has different requirements, the engineer manager will have an opportunity to develop the type of information that he requires.

Another important aspect in the design of the information system is finding the best source. The purpose is to have accurate information available quickly. This requires the use of the best information source for the area of need. Sources include employees, business consultants, labor union officials, legal counsel, and company banking relationships.

The engineer manager may also be concerned with establishing a realistic effective communication technique from source to user. There are many different information types and as many ways to transmit the information. The problem is determining which one will accomplish the design objective at the least cost. The spoken word often becomes the most effective because certain words, tones, pronunciation, and gestures can be used to facilitate communication. They provide a two-way flow which is especially useful in trying to transmit difficult information. Face-to-face communication involves considerable time, but it seems to be the only way complex information can be communicated accurately.

The engineer manager has available other means of transmitting information which are effective, fast, and cheap. Television, telephone, and radio are effective communication means that he should always consider in the information system design. The printed word is especially effective when information is factual and rather easily grasped.

The manner in which an engineer manager communicates says a great deal about him and about his sense of logic, accuracy, and clarity. Every manager has a style of communication that has its strong and weak points. He must be conscious of these and emphasize the good ones and improve the weak ones. The following suggestions can serve as a helpful checklist for improving communication style:

DEVELOPING AN APPROPRIATE STYLE

1. *Data and words.* Avoid abstract words and data that tend to be vague or ambiguous. Use data and words that are clear, specific, and to the point.

2. *Statements.* All statements of information should be short. In most cases, shorter statements are more readable than long involved ones.

3. *Varied information.* Use a variety of statements and types of data in the communication process, e.g., data, figures, and graphs. In asking questions, use imperative sentences because they make the message more interesting.

4. *One idea.* Have each report, message, or communication in the information system concerned only with one idea.

5. *Directness.* Be direct and specific to the individual or group; for instance, request the plan from Paul or the chairman of the committee.

6. *Unnecessary items.* Do not use unnecessary words, data, or information.

7. *Emphasis.* Use voice, spacing, underlining, repetition, charts, and color to emphasize important information.

8. *Feedback.* Plan to ensure that people understand the communication by using questions and answers, requesting them to restate in their own words, using questionnaires, or observing reactions.

9. *Follow-up.* Since a vital part of communication is follow-up, use it to reinforce the communication. Highlights, progress reports, confirmation briefings, interviews, memos, or telephone calls can be used as ways to follow-up.

The engineer manager must develop his own style of communication, one with which he is comfortable and feels natural. His style will vary from situation to situation, but basically he will soon find the style that is most effective for him.

In addition, his style will change as he becomes more competent with certain techniques. As he gains confidence in computer and data processing, he will use this information system more often. The same with speaking; at first he will write his speeches out word for word, and they will usually sound stuffy and unnatural. But later, as he gains confidence through practice, he will only make notes and then he may not even use them.

DEVELOPING EFFECTIVE SKILLS

The engineer manager, in developing the information system for his organization, must develop effective skills to fit not only his personality but also the group's personality. He must realize this at the initial stage of developing his information system.

The function of the system is to provide information to individuals who need the information for guidance in their actions. The system also fulfills people's desires for awareness of things that affect them. The manager will find that employees are hungry for information, generally do not get enough, and are always interested in information related to their job. People are interested in:

1. How their work fits into the goals or objectives of the organization
2. Chances for advancement
3. Outlook for the firm's product
4. Prospects for work security
5. Profits and losses, income and expense
6. What is expected of them and how well they are doing
7. Reasons for changes
8. New products
9. New policies and even old ones
10. History of the company
11. Where products are sold and how they are made

People are very interested in learning about the actual operations of the organization they work for. The manager must continue to develop skills that will improve the dissemination of such information. Some of the most important skills that he must acquire if he is going to be effective in this are set forth below. The engineer manager must be the following:

1. *Noticer.* The engineer manager must be an expert in noticing how people are reacting to his communication process. People express themselves in many ways, such as absence of reply, folded arms, fidgety motions, or drooping eyelids. He must recognize that these communication techniques are reciprocal participation in his communication. He must have the skill to recognize these "patterns" that

less careful people may overlook. He must continually search for things that indicate how his messages are being received. Even after the communication event, he must still recognize and notice how well his message was received and how its delivery can be improved.

2. *Reader*. The engineer manager must acquire efficient reading habits. He is faced with a mountain of communication that reaches his desk from both internal and external sources. He must sort out those things important to transmit to his subordinates. At the same time he must maintain currency in his profession. To achieve the reading skill necessary to accomplish this, he should take advantage of any rapid reading clinics available to him. And then he must practice the techniques he learns. The engineer manager can make some improvement in his reading skills by doing the following:

A. Make a preliminary examination of different reading matter as follows:

   (1) *Books:* Examine title, author's name, publisher's name, date of publication, introduction, and preface. Look over table of contents since it is an outline of the book. Thumb through quickly. Read glossaries and summaries.

   (2) *Reports:* Examine title, author's name, company, and department. Check date, for whom prepared, purpose, table of contents. Read summary and abstract. Note headings and graphics. Read conclusions and recommendations.

   (3) *Letters:* Examine letterhead, date, name and title of writer. Check main idea.

   (4) *Magazine and journal articles:* Examine title, author's name, biographical notes. Note headings, subheadings, and graphics. Examine conclusions and summary.

B. Read material in phrases. Recognition span should be widened to include several words to reduce the number of fixations and thus save time.

C. Concentrate. The greatest impediment to efficient and retentive reading is inattention. The mind must be fully engaged.

D. Practice. The goal is to quicken the perception rate so that the mind recognizes and accepts images from the eyes.

E. Learn to skim. Skip from one important part to another. Skim rapidly to find a fact.

F. Build vocabulary. One cannot read rapidly unless the mind recognizes the words and phrases instantly.

G. Select appropriate reading speed. A news magazine can be read and understood more rapidly than a technical article.

3. *Evaluator*. The engineer manager must develop the skill to identify elements in an orderly manner, i.e., give order to the thoughts perceived. Through this skill he identifies the essential features of the message.

4. *Organizer.* Most information comes to the engineer manager in miscellaneous parts that must be combined into a pattern or organized system. Through organizing he is able to link the information to other available knowledge and simplify interpretation. The manager's skill in organizing communications will affect the quality of his messages.

5. *Note taker.* The engineer manager must develop the skill of recording information in an orderly, organized manner. He should use a loose-leaf notebook and arrange his notes so that he can find them quickly later. He should use a numbering system and date each page. All notes should be in his own words and stated as briefly and concisely as possible. He should practice taking notes and, if necessary, reword them later.

EFFECTIVE LISTENING Although a large part of every engineer manager's time is spent listening, generally he is very inefficient in this skill. Since listening provides him with a large portion of his information, listening is very important. It means tuning in to the speaker's feelings, attitude, and frame of mind. Because every engineer manager must listen effectively, he should use the techniques listed below to sharpen his listening skills:

1. Listen positively to everyone. Be receptive. Have an open mind.

2. Keep quiet and do not interrupt, except to give a word of assurance. Allow the speaker to make his point before reacting.

3. Pay close attention to the speaker's words.

4. Analyze the speaker's attitude. What is his frame of mind? What kind of person is he? A pessimist? An optimist? An alarmist? Unpredictable? Recognize the feelings and emotions in the speaker's message.

5. Listen beyond the speaker's words. Ask "What does he mean by what he says?" The words themselves often count less than the intent behind them.

6. Do not make an immediate value judgment. The speaker's apparent meaning may change as he develops his thought and position.

7. Do not mentally argue with the speaker.

8. Understand what the speaker means by his terms and words. Different meanings of words frequently lead to misinterpretation of ideas or positions.

9. Avoid wishful listening. Many times the engineer manager hears what he wants to hear, not what the speaker says or means. One way to check this is for the engineer manager to repeat the idea to the speaker in his own words, such as "Do I understand that we will be able to purchase the new machine?"

10. Increase memory span by repeating words and numbers and by writing down the events of the day.

11. Watch the speaker's body language to see if his physical actions and facial expressions agree with his messages. If the speaker appears uneasy, uncomfortable, or evasive, then the full story is not being presented. An individual who is uncommitted to an idea usually shows this unless he is very skillful at deception.

The technique of listening is one communication means that the engineer manager can always improve. Listening requires separate attention and cannot be done simultaneously with something else. Speaking rates are about 150 words per minute; listening rates are over 1,000 words per minute. The idle time must be used to review and assimilate what is being said. To improve his listening skill, the engineer manager should periodically check to see if he is practicing effective listening techniques. He should ask himself if he is:

1. Interpreting facial expressions

2. Interpreting body language

3. Able to understand the intent of the words

4. Searching for the meaning of the speaker's words

5. Listening for the real meaning

6. Avoiding wishful listening

7. Evaluating how the speaker says things rather than what he says

8. Giving attention to the source of information, experience, observation, and prestige of the individual

9. Determining the speaker's assumptions

10. Separating out the facts from the opinions of the speaker

11. Trying to avoid misunderstanding

12. Accurately restating the speaker's ideas in his own words

13. Trying to determine the speaker's thoughts, questions, goals, or objectives

To the engineer manager, listening can be interesting and exciting if he acquires the skills of listening. Otherwise, he will consider it a waste of time, and will thereby miss out on much information that he might otherwise have available to share with his subordinates. Even worse, if his employees are the senders, his poor listening habits will turn them off and they will cease to provide him with important information that may affect his ability to perform. Effective listening can influence all functions of the organization.

The engineer manager can improve his communication through effective use of words, both written and spoken. He must develop unity, coherence, and emphasis in his communications.

The engineer manager must acquire *unity* in his writing and speaking, which means conveying one main idea in his message. Unity requires that a sentence convey only one idea or modifying ideas which may complement the main idea. He may unify his sentences by avoiding fragments, loose connections, and word omissions. These are common problems in both writing and speaking. At times he may write his messages in telegram form in which the reader must supply missing words, but to avoid misunderstanding, this form should be used infrequently. He must include every word or number or piece of information necessary for his message to be understood completely and quickly.

The tying together of sentences or thoughts is called *coherence.* To achieve coherence, words such as "however," "moreover," and "consequently" or short phrases are used. Another method to improve coherence is to keep the grammatical person, number, and voice consistent. When ideas are similar, this consistency aids the reader in recognizing their similarity. Shifts in sentences and thoughts must be avoided and parallelism of words, phrases, clauses, and sentences should be used.

To make messages more meaningful, the speaker must use *emphasis.* This requires effective word order, deliberate climax, contrast, conciseness, and correct sentence structure. Since the beginning and ending of a sentence attracts special attention, the key idea is put at the beginning or the end rather than in the middle of the sentence. A useful technique to gain emphasis is to arrange words, phrases, and clauses in a series of increasing importance so that each thought or idea supports one major thought. In most cases, the main idea is postponed until the end along with other phrases and words used to stimulate and support the main idea.

As the engineer manager's staff grows, he no longer has enough time to speak daily with everyone. Therefore, instructions, policies, and procedures must be transmitted in written form. The following techniques will assist in sending clear messages:

1. Consider the people involved. The engineer manager should adapt his message to the feelings and ideas of the individual or group receiving the message. He should accommodate the individual's personality so that words and phrases inform or persuade the reader or listener. The individual's mental and emotional state, formal education, training, experience, needs, ambitions, and desires should be considered in developing his messages.

2. Determine how and when the message should be sent. For the engineer manager, the how and when are very important considerations when sending messages. The objective is for someone to accept what is sent by writing or speaking.

At times, certain messages are inappropriate. At other times, a written message is more effective than an oral one.

3. Emphasize the most significant idea or thought. Whether for the listener or for the reader, deliberate devices are used to direct attention to the message. The most common devices for writing are capitalizing, punctuating, listing, underlining, and spacing and the most common devices for speaking are pauses, facial expressions, gestures, volume, and rate of speaking.

4. Choose the correct word. The engineer manager should use words and phrases that the receiver understands. He should choose words that help the reader to see the idea in relation to his experience and that help him to emotionally understand the idea. He should not use words that are uncommon to those receiving the message. He must remember that the goal of the transmission is to provide a clear, understandable message.

5. Use traditional spelling, punctuation, and grammar. Poor grammar obscures the message and cheapens the engineer manager's reputation. He should consult a current dictionary when in doubt about the spelling or meaning of a word. He should use a thesaurus to seek alternate words that may clarify the meaning for the receiver.

6. Avoid needlessly wordy phrases. An effective method to improve communication is to develop concise phrases. Sometimes the engineer manager is tempted to use long, involved phrases and statements to conceal his lack of knowledge of a subject. This obscures the meaning of the transmission and the effectiveness of the message.

7. Communicate with courtesy in mind. To achieve courtesy, the engineer manager must consider the needs and desires of his readers above his own. Since most people resent rudeness, indiscretion, and deception, he should avoid any evidence of these in his communication.

Effective communication comes easily to engineer managers who learn and practice using communication techniques. Mastering these skills and techniques will take time and effort, but the benefits received will far outweigh the effort. Engineers have long been condemned for their inability to communicate. If they are to become effective in the role of engineer manager, the mastering of such techniques is essential.

The engineer manager spends approximately 90% of his time   *SUMMARY*
communicating in one form or another. He must effectively use
a communication process that involves his transmission of a
message which is encoded, put through a particular channel,
decoded by the receiver, and then fed back to him to assure that
the message has been received. For his message to be effective, he must know the background of the people, their expectations, their education, and their concern for their personal welfare.

There are a number of communication or information systems, including electronic data, formal, verbal, technical, and informal. The environment within the organization will have an effect on successful communication of information. There are a number of conditions that cause subordinates to be unwilling to communicate with the manager and he must be aware of certain conditions that will make the message better received.

There are several positive steps that the engineer manager can follow to encourage improvement in information transfer, including clearly expressing the message, interpreting the message correctly, developing trust, listening, and keeping an open mind.

The information system should be designed along organizational lines if it is to be an integral part of the organization. The information system is a vital link in allowing the achievement of goals or objectives.

The manner in which an engineer manager communicates has a great effect on his superiors and subordinates. Each manager must develop his own style, one with which he is comfortable and with which he feels natural. He must be aware of the interests of his people and endeavor to keep them properly informed.

The engineer manager can develop a number of techniques that will be useful to him as a communicator. An important technique is developing his reading abilities. Another is developing the habit of effective listening.

Finally, the engineer manager must improve his communication skills through the effective use of words, both written and spoken. His messages should have unity, be coherent, and have the proper emphasis. His future depends on his ability to communicate.

Important Terms

Communication Process:	A system for transmission of a message from sender requiring encoding, delivery, decoding by the receiver, and feedback to the sender.
Effective Listening:	The technique the listener uses to secure the greatest amount of information transfer from the speaker. The technique includes ample review and assimilation.
Formal Information System:	Formal chains of command in any organization through which official information must flow.
Informal Communication System:	A communication system often referred to as the "grapevine." It is generally structureless and goes into operation when members

of the formal organization who know each other well begin to pass on information dealing with the enterprise.

Style of Communication: Each manager must develop a style of communication with which he is comfortable and with which he feels natural. His style will evidence itself in both the written and spoken word.

For Discussion

1. Discuss why effective communication is important to the engineer manager.
2. Define the communication process. Why is feedback important? To understand the communication process, what must the engineer manager know about the receiver of the message?
3. Name five specific concepts that are important to the information process.
4. Name four types of information systems. What is involved in the formal information system?
5. Discuss the operation of a "grapevine." Why is it important to the manager?
6. Name five problems encountered in developing an information system.
7. Name five conditions that cause subordinates to be unwilling to communicate with the manager. What can the manager do to correct this?
8. Name six important steps that lead to an effective information system.
9. Discuss the designing of an information system.
10. Discuss what is meant by the manager's style of communication.
11. Name seven things that seem to interest employees.
12. Name five skills that the manager should acquire if he is to be a good communicator.
13. Outline some of the techniques that will improve reading skill.
14. List and discuss eight techniques leading to effective listening.
15. What constitutes effective communication? Outline the elements of effective oral communication and effective written communication.

Case 11–1

Implementing a New Program

R&D was requested by Department XL-19 to study ways and means to increase productivity and reduce the costs of its operations. Rex Smith formed a project team to study the problem for the department.

The first month the group spent getting familiar with the department's operations and its personnel. Most of the personnel were reluctant to talk until they found the engineers had no axe to grind. It was a pleasant surprise, as one Department XL-19 employee stated, to see that the team was more interested in listening than telling. The engineers gave no evidence that they were going to change things. Some of the employees even wondered what good they were doing. During the next month the engineers made very few visits to Department XL-19.

The following month the team put together a program that would increase output and do it with less labor. Rex and the engineers met with Hernandez Garcia, head of Department XL-19, who agreed to the program with minor changes. Then Rex and the team, along with Garcia, met with the department supervisors and carefully explained the program. They pointed out that each supervisor would be expected to work with his people to implement the program.

After the meeting, two of the supervisors were overheard talking about the program. One of them said, "Well, after we told them all we knew, they give us this unworkable program." The other supervisor agreed, "Yes, and it will be a long time before they get anything more out of me. I will make sure it will not work."

1. What do you think of the supervisors' reactions?
2. What approach should Rex and his group have taken?
3. What could Garcia have done to aid in the acceptance?
4. What can be done now to implement the changes?

Case 11–2

Too Much Pressure?

Tom Johnson, Vice President, Metal Products Company, was reviewing the past year's performance and some data from Market Research on their market share. He noted that the product lines from Department XL-19 were losing some of their market share. Also, in comparing prices, he noted that XL-19 products seemed to be priced higher than those of competitors.

Tom called Hernandez Garcia, the head of Department of XL-19, to his office and showed him the figures. He told him that he had to get the costs of his products down by 10% so that they could be competitive and get back their share of the market.

Garcia took the problem seriously. He called in all of his supervisors and told them what had to be done. He urged them to get a little extra out of their workers. They were to speed up machines and assembly lines, reduce waste, and cut down on rest periods.

Some bad things resulted. Absenteeism and employee turnover increased. The supervisors complained about the workers' attitude.

1. In what way is this a communication problem?

2. What specific steps should Tom and Garcia take to correct what seems to be happening? Why?

For Further Reading

ANASTASI, T. E., JR., *How to Manage Your Reading*. Cambridge, Mass.: Executive Development Center, 1965.

BURACK, A. S., *The Writers Handbook*. Boston: The Writer, 1972.

DAVIS, K., "Human Behavior at Work," in *Organizational Behavior: A Book of Readings,* 5th ed., ed. K. Davis. New York: McGraw-Hill Book Company, 1977.

DUBIN, R., *Human Relations in Administration,* 4th ed. Englewood Cliffs, N.J.: Prentice-Hall, Inc., 1974.

ELLIS, D. S., *Management and Administrative Communication*. New York: Macmillan Publishing Co., Inc., 1978.

HALL, J., "Communication Revisited," *California Management Review* (Spring 1973), 56–67.

HANEY, WILLIAM V., *Communication and Organizational Behavior,* 3rd ed. Homewood, Ill.: Richard D. Irwin, Inc., 1973.

HARRIMAN, B., "Up and Down the Communication Ladder," *Harvard Business Review* (September 1974), 143–51.

HODGES, J. C., and M. E. WITTEN, *Harbrace College Handbook*. New York: Harcourt Brace Jovanovich, Inc., 1977.

KATZ, D., and R. KAHN, *The Social Psychology of Organizations*. New York: John Wiley & Sons, Inc., 1978.

KEEFE, W., *Open Minds: The Forgotten Side of Communication*. Belmont, Calif.: Wadsworth Publishing Co., Inc., 1970.

LINDAUER, J. S., *Communicating in Business*. New York: Macmillan Publishing Co., Inc., 1974.

MINTZBERG, H., "The Manager's Job: Folklore and Fact," *Harvard Business Review* (July–August 1975), 49–61.

RICHARDS, M. D., and W. A. NIELANDER, eds., *Readings in Management,* 4th ed. Cincinnati: South-Western Publishing Company, 1974.

ROCKEY, E. H.,*Communicating in Organizations*. Cambridge, Mass.: Winthrop Publishers, Inc., 1977.

SCANNELL, EDWARD E., *Communications for Leadership*. New York: McGraw-Hill Book Company, 1970.

SCHAILL, W. S., *Read Faster, Read Better for Pleasure and Profit*. New York: The Developmental Research Institute, 1962.

SHURTER, R. L., *Effective Letters in Business,* 2nd ed. New York: McGraw-Hill Book Company, 1954.

TUBBS, S. L., and S. MOSS, *Human Communication*. New York: Random House, Inc., 1974.

WICKESBURG, A. K., "Communication Networks in the Business Organization Structure," *Academy of Management General* (September 1968), 253–62.

PRESENTING PROPOSALS EFFECTIVELY

The engineer manager must effectively present proposals and ideas to his subordinates and superiors. To accomplish this, he must know what information is important and the proper techniques for presenting it. He must train the engineers in his group to use the techniques effectively.

LEARNING
OBJECTIVES

☐ Learn how best to present ideas to superiors and subordinates.

☐ Determine ways to create a strategy for making a presentation and assessing the group.

☐ Learn how to prepare the presentation.

☐ Develop methods for making the presentation effective.

☐ Gain insight into how to be effective at meetings.

EXECUTIVE COMMENT

PAUL T. DOWLING
Chairman and Chief Executive Officer
Nooter Corporation

After experience with Jones and Laughlin Steel Company and Granite City Steel Company as a metallurgist, Paul T. Dowling joined Nooter Corporation as sales engineer. He moved up to divisional sales manager and general sales manager, vice president, executive vice president, president, and then to his present position as chairman of the board.

Presenting Ideas Effectively

Traditionally, the greatest weakness of the typical engineer has been the inability to communicate effectively with others, especially with non-engineers or nontechnical people.

The best idea in all creation is worthless unless it can be understandably presented to the person or group of persons in a position to authorize its continued development. The idea must be "sold," if you will. The engineer must empathize with his audience, fully relate to his audience, and present his thoughts and ideas in terms that are totally familiar to and readily understood by that audience.

It is hoped that a desire will be kindled within the engineer manager to develop a facility for expressing his ideas, both orally and in writing, in the idiom of the expected audience.

In the preceding chapter we studied the general principles of effective communication within the organization. We will now turn our attention to more formal presentations to superiors and subordinates.

The alert manager begins to see predictable patterns of behavior developing in his group. As he works with people, he is able to anticipate their reactions. For example, he anticipates objections and develops a response beforehand. He presents his case in such a manner that the response to potential objections has already been made.

After the engineer manager has successfully presented his case to the group and it appears to have acceptance, it is wise for him to follow up to ensure that his ideas are being carried out. This follow-up tells them that the matter is important to him. It has the additional advantage of getting him out of the office and in contact with his subordinates.

Before the engineer manager can effectively present his ideas and influence others, he must first clearly define the goals of his group's activities. Frequently an engineer is criticized for an action that is not compatible with the group goals. This failure may well be because he is not aware of what his group is supposed to achieve. Communication between the manager and his group cannot be effective until both know where they are headed, i.e., they both understand the goals.

KNOWING THE PURPOSE AND GOAL

Presenting Ideas to Superiors

In every meeting he has with an individual or group, the engineer manager hopes to achieve something. For example, the engineer manager is preparing to meet with the top management of his organization to request approval to purchase $500,000 worth of new equipment. His ultimate objective is to secure approval of the $500,000 equipment purchase, but his immediate objective is to influence those attending the meeting to respond affirmatively. What do they have in common? He knows from their backgrounds that of the ten attending, only three will be from production and have engineering backgrounds. The others will have sales, accounting, and legal backgrounds. Also, all of them have specific responsibilities in the company and will look at his request in the light of these responsibilities. But they do have one thing in common: They are all interested in the profitability of the company. Now the engineer manager's objective becomes clear and specific. It is to convince them that the equipment will increase the productivity of the operation and in turn reduce cost per unit and increase the company profit.

In any presentation the engineer manager should limit himself to one objective. Most people believe they should use all the allotted time and thereby try to accomplish too much in a meeting. As a result, they end up accomplishing little because they try to convince their audience of too many different things at one time. To be effective, the engineer manager must devote considerable time and effort to defining his specific objective. It is also important that he rehearse his presentation, preferably in front of colleagues. An excellent way to prepare is to give the presentation before a TV camera and then critique the results. Since selling ideas and proposals is a major responsibility of the manager, too much time cannot be spent in preparation.

The engineer manager is and must be the primary source of knowledge and information about the subject being presented. There is a great deal more, however, to being well informed than merely being able to answer questions during and after the presentation. The ability to influence people goes beyond supplying information. People react to the engineer manager on the basis of their confidence in him. Knowledge of the subject gives the engineer manager confidence in himself and his ability to share it with others. In addition, knowledge and confidence make him enthusiastic about his proposal. This confidence and enthusiasm are automatically passed on to his audience. His enthusiasm increases the effectiveness of his communication skills and sows the seed for better understanding by those who hear him. As with the common cold, enthusiasm is contagious.

In preparing for a presentation, the manager must carefully review each situation, the kind of people who will be present, and his objective. This will help him to determine the information that is most essential. But in determining the exact facts and knowledge required, it is best that he carefully examine the scope of the goal being considered. No engineer manager works in an isolated area, for no idea and no program can stand by itself. It is related to and will affect the whole organization, and many times it will be affected by the industry and the economy. The engineer manager must clearly understand how his objective fits in with broader corporate objectives. Only in this way can he appeal to the biases of those who will be present.

Presenting Ideas to Subordinates

Most ideas or proposals, especially complicated ones, are not judged strictly on their own merits. The qualifications of the manager and the respect his audience has for him will influence the audience's receptivity. A comparison might be drawn to brand names. A well-known product brand name is much easier to sell than an unknown product brand name. Therefore, the engineer manager must be very conscious of his own qualifications and strengthen them through his knowledge of the subject. If he can produce evidence from outside or inside the corporation on the validity of his proposal, he will enhance his chances of selling it.

Many managers fail to have adequate knowledge of what is going on in other firms when they try to sell something to their own group. If an engineer manager is trying to sell his people on a new wage and salary policy, he needs to know what is happening in other firms in his vicinity and what is happening in the industry. He must have adequate knowledge if he is to build a case that is convincing enough to get the group to accept his ideas. He must know the strengths and weaknesses of his proposal and he must be equally well prepared to discuss alternatives. Only when the engineer manager is armed with such knowledge can he talk and act with real authority and confidence.

The reasons people are frequently unwilling to accept the engineer manager's ideas are that they lack trust in him and they fear that he will be unable to make good his promises. Therefore, a record of not carrying through with past promises can make future selling jobs all but impossible. This is why people prefer to listen to and accept advice from managers who have a record of reliability. Regardless of whose fault it is, the engineer manager is the one who will be held directly responsible and will receive the blame if things do not work out as he has presented them.

When an engineer manager presents an idea to his group, he should back up what he says with action. He should anticipate questions from the group and answer them effectively. He recommends how to overcome problems. He keeps everyone informed. He is able to foresee possible stumbling blocks that may affect achieving his goal. He has planned for the human and material resources required to achieve his goal. His answers are specific and direct. He is able to put his group at ease.

The engineer manager must have a thorough knowledge of the goal that he is presenting because (1) ideas and solutions to problems are becoming more complicated and sophisticated and (2) the people he is trying to convince are better educated and more demanding. He needs to know:

1. What his goal (idea) can and cannot do
2. How his proposal compares with the present situation and other alternatives
3. How his idea will solve the problem as viewed by his audience
4. How his idea will affect other parts of the organization
5. What other advantages are related to his idea

In answering the above questions, the engineer manager must learn to deal in documented facts. In the case of the wage and salary policy mentioned above, if he were asked how it has worked, the manager might answer, "It is working fine." But if he were to respond that in a study by the ABC research firm it was reported that the system was used by 35 companies listed in the Fortune 500 and that it had been in use for over 5 years by 30 of these companies, his audience would be more impressed. If he had a copy of the study to present to them, his credibility would be further increased. Frank admissions about the proposed idea and what it can or cannot do always build confidence.

Knowledge about alternatives gives the manager making a presentation the upper hand. The people he is talking to may be knowledgeable about these alternatives, since others may have discussed similar alternatives. Lack of knowledge about alternatives seriously undermines confidence and makes the audience wonder how thoroughly the manager has investigated the subject. The proposal may not be the best solution, but a knowledge of alternatives boosts the manager's confidence in himself and the group's confidence in him. Gaining knowledge about various subjects is an endless job which involves careful reading, visiting with other units, requesting information from staff, going to conventions, visiting with salesmen, and going to professional meetings. Developing good, effective sources of information is important if the engineer manager is to keep well informed. His ability to do this depends on the rapport that he has developed with others. His objective should be to develop a working relationship with people at all levels in the organization so that when it is necessary he is able to obtain information or opinion.

CREATING   Once the engineer manager has defined his presentation goal,
A STRATEGY   his next step is to plan the method to be used to tell his story, i.e., develop a strategy. If his presentation is designed to appeal to individual personal interests or the perceived group interests, his chances for having his ideas accepted are greatly increased.

Presenting ideas is similar to selling. The presenter wants the group or an individual to buy the idea or proposal. Therefore, the engineer manager must use a strategy that indicates that his goal or idea is in the best interests of the group. Although people's interests vary greatly, there are several common interests, for example:

1. Financial: profitability, increased productivity, lower cost, increased salary and wages, security
2. Human: fear of labor-saving devices, effort, reliability, usefulness
3. Pride: self-respect, acceptance by group, popularity, rank, position, possessions
4. Ethical: trust, fairness, conscience, religion, sympathy

The engineer manager must determine the specific interests of the group members and the degree of importance they attach to these items. He must carefully analyze his group or the individual he is talking to and appeal to the most powerful interest factor. Determining the greatest interest of a group or individual is difficult, however, because of the dynamic characteristics of individuals. He should analyze each situation from the standpoint of the following:

1. *Likes* and *dislikes* of the members of the group. This is based on personal knowledge gained through working with the people, meeting with them, and learning to recognize their interests.

2. *Characteristics* of the people that will help determine interests. For example, education and age give clues to common interests and relationships.

3. *Position* in the organization. Engineers have special interests as compared to accountants or lawyers, middle management, or top management.

4. *Background of the people.* Regions or areas where people were raised and their religious backgrounds indicate their interests.

The engineer manager can use this knowledge of the common interests of people throughout his presentation, for example:

1. In his opening remarks
2. By proving his points by using examples with which they are familiar.
3. By talking on the appropriate level
4. By using terms familiar to the group
5. In reaching an agreement

To appeal to people, the engineer manager must speak in terms of the group's interests. One way is to use the word "you" at every possible opportunity. This may require his rephrasing a sentence or thoughts, but it will make people feel a part of the presentation—and this is important.

During the presentation the engineer manager wants to "get through" to his audience. To do this effectively, he must assess their reaction and the way they think. People tend to run fairly true to type in the clothes they wear, the cars they drive, the people they associate with, the way they work, and the way they relate to other people. An individual has a "style" that dominates his behavior. The manager must understand this style of behavior and tailor his presentation accordingly.

ASSESSING THE GROUP

Knowing an individual's style gives the manager some insight into how the person thinks, the kinds of activities he enjoys, and his priorities in life. If the manager does not know the individual's style of behavior, he may actually offend the person without intending to, which, of course, can be disastrous when the goal is to sell an idea or proposal.

There are many different styles of behavior. These behavior styles are grouped into the following four broad classes:

1. *Detailers.* These are people who pay close attention to "all" details. They always check and recheck their work to be sure that it is correct and letter-perfect in every detail. They love to get involved in research and they collect all the facts before making a decision. These people are rather conservative in both dress and appearance at the workplace and at home. Detailers are generally the engineers, computer scientists, lawyers, and physical scientists.

2. *Affecters.* Affecters love everyone. They need to be involved all the time in order to feel secure in everyday life. They always need more friends and they have a strong desire to make friends. They are warm people, always on the move, and always wanting to try something new. But they have a difficult time following through and completing a job. Affecters need to be involved. In fact, one of their problems is that they cannot say "no," consequently, they soon become over-committed. These people are generally in sales, social work, acting, and nursing.

3. *Strangers.* These people are sometimes considered odd. They are rigid in their thinking and tend to become very impatient when other people do not understand the value of their ideas. They are very imaginative and they are rather impractical in their approach to everyday problems. This group consists of the inventors, artists, and architects.

4. *Pushers.* Pushers are the most common type of people in our society. Their behavior is goal-oriented and they are always trying to achieve these goals. Hard work and a quick reward are important to these people. They give full loyalty but they also demand the same. This group is made up of managers, athletes, and politicians.[1]

For the engineer manager to be successful, he must (1) realize and understand his own behavior, (2) develop the ability to recognize other people's behavior styles, and (3) respond to these different behavior styles in the group. This may seem like an impossible task to the new engineer manager but clues are everywhere, and he must develop the habit of looking for them. He should observe how people talk to him over the telephone and how they keep their offices and workplaces. For example, in general, pushers are unconcerned about the appearance of their offices; detailers have clean, neat offices; affecters have personal momentos and pictures in their offices; and strangers are erratic.

When the engineer manager is able to determine the other individual's behavior style, he can make his presentation with that person's particular style in mind rather than his own. He will use words that tend to fit the other person's behavior style rather than his own. Words and phrases that fit the behavior styles listed above might be something like this:

1. Detailer: cautious, specific, evaluate and analyze, astute, well planned, follow plan, thinking through, critical

[1]Paul Mok with Dudley Lynch, "Easy New Way to Get Your Way," *The Reader's Digest* (March 1978), pp. 105–09.

2. Affecter: wow, sensitive, existing, emotional phases, stimulating involvement, meaningful actions, help a friend, carried away
3. Stranger: future ideas, relate ideas to each other, new, idealistic, creative
4. Pusher: urgent, today, now, results, practical, react immediately, get things done, today counts, must justify activity

If the engineer manager presents his ideas in a language that fits the behavior style of the person or persons he is trying to influence, he is more likely to achieve his goal. But more important, by using such an approach in evaluating behavior styles, the engineer manager is always aware that people look at an idea or proposal differently. He must present it so that it is appealing from their perspective.

The manager's chances of success in turning, changing, or influencing an individual or group to accept his idea or proposal increase according to the degree of homework he has done. Effective preparation establishes the stage for and immeasurably facilitates success. For the successful engineer manager, preparation means:

PREPARATION OF THE PRESENTATION

1. Providing himself with adequate knowledge of the subject
2. Providing adequate support, i.e., notes, visuals, records, and data
3. Maintaining good physical condition through adequate sleep and exercise
4. Disciplining himself to set a clear goal for the meeting
5. Reviewing reports appropriate to the case

Interest and participation of the group are at the core of any good presentation. But the engineer manager must go further than just planning the content and support media for the presentation. He must also be concerned about his dress, the words he uses, and the manner of approaching his subject. He will use an entirely different approach for a group of engineers than he will for a group of production line workers. He would probably address the production line workers in shirtsleeves and open collar and talk in simple, homey terms they can understand. He would probably want to speak to them in the early part of their work shift. For a group of engineers, he would be more formally dressed and speak in terms familiar to the engineers and would perhaps meet them later in the day. For his superiors, he would be more formally dressed, have well-prepared visual aids, and use words and phrases that top management uses. For example, the general foreman and plant manager speak in terms of cost of manufacture, the general manager speaks in terms of dollars of profit or return on the investment, and the president and chairman of the board speak in terms of earnings per share of stock.

The engineer manager must base his organization of the presentation on the reactions of the listener(s) or the group. This helps him to anticipate the group's objections and provide a means of overcoming them. Frequently, the first reaction to the presentation that the manager will sense is apathy. A few may really be interested in hearing what he has to say, but for the most part, he may feel that the group is uninterested in his remarks, ideas, and thoughts. To overcome this, the manager should open his presentation with a statement that will immediately catch attention and interest, for example, by hinting that something important is going to come, or by using a well-directed challenge, a well-told anecdote, a quotation or saying, or even a question (but the manager should be prepared for someone to answer it). These opening remarks are designed to change the group's disinterest into interest. He must arouse his listeners' interest and this requires a thorough knowledge of his audience.

His next step is to maintain the group's interest by informing them about how his proposal will affect them *"personally."* He must tell the group why they should be interested and how they will benefit from it. He may appeal to their pride and their desire to overcome difficult problems.

The third step is to present his proposal in detail and give supporting facts. He must avoid making general statements. What is needed now is convincing information—statistics, examples, quotations from others—to remove all doubts from the group. For this phase, he needs to use illustrations and examples to support his points. By using illustrations and examples to support his points, the engineer manager is restating his points and taking advantage of the effect of repetition. The format would be as follows:

1. Present the thought.
2. Give illustrations and examples using statistics and other supporting evidence.
3. Repeat and use additional illustrations and examples.

It has been said that we remember only 10% of what we hear but 50% of what we see and hear. The use of visuals helps the manager to achieve the 50% level.

To reiterate, the most common mistake managers make in presentations is to try to cover too many topics. If it is necessary to cover more than one topic, each topic should be handled as described above and in a clear and precise manner.

The last phase of the presentation is the "conclusion," which will include a request for approval or for action. (It is similar to a politician's asking the audience to vote for him at the end of his speech.) If the presentation is to the manager's superiors, it will likely be a proposal for approval. If the presentation is to his subordinates, it may be an idea for adoption or a request for action on a decision that he has made. There is an old cliché that says, "At the beginning of your presentation tell them what you are going to tell them, then tell them, and in conclusion tell them what you told them." The cliché has some merit if it is done in good taste and is not too obvious.

The engineer manager should not start a meeting or presentation without first determining the demands that will be made of him. He should visualize each step of the presentation. He should ask himself what he will need to be able to do to meet all the possible circumstances that may arise. In making the presentation, the manager will be required to use both mental and physical resources. The mental resources will involve his verbal presentation and the physical resources will involve visual aids. Visual aids can be tailored to appeal to a specific group. They represent a major part of any presentation. As a Chinese proverb says, "One picture is worth more than ten thousand words." Therefore, visual aids make ideas easier for the group to understand. The engineer manager must examine his presentation to determine which parts need visual assistance and which parts can be told more effectively.

When the engineer manager reviews his presentation, he should look for areas that are unclear or difficult to express. For those, he should prepare visual aids that will make the points understandable to his group. He must decide whether to use professionally produced visuals or his own homemade visuals. In most cases, the personalized homemade visuals are the most effective; but in making major presentations to superiors, more professionally prepared visuals may be desirable. In most major corporations, professional assistance is available from the public relations, advertising, or sales departments. They are helpful in preparing the most appropriate visual aids, for example, 35mm slides, transparencies, motion pictures, or flipcharts. The manager will have to give them the basic data and an idea of what he wants to convey.

Another effective visual aid is the actual part, the item, the model, or the building. Taking members of the group on a plant tour allows them to use all of their senses as a means of understanding. The engineer manager is limited only by his imagination. Often the most simple visual aids are the most effective.

Although visual aids are a very helpful tool for presenting ideas, too many visual aids that do not fit the subject may detract from the presentation. Visual aids should supplement and complement the presentation so that it may be understood by the audience.

Developing the Presentation

In preparing the presentation, the manager plans for alternatives and emergencies. He does not rely on the inspiration of the moment. He should always spend adequate time in planning the presentation. He must constantly ask himself the question, "How can I make this presentation influence the minds and emotions of this group?" In developing his presentation he should:

1. Have a thorough knowledge of his goal
2. Analyze the individual needs of the group members
3. Help the group to recognize its needs
4. Be able to meet those needs

If the manager looks at his ideas and method of presentation in this manner, he may realize why many of his ideas at times seem to be unacceptable to his people.

The engineer manager must learn how to use the correct words in his presentation. He must use all of his words with skill and thought and he must use the right word at the right time. Words may have different meanings to different groups of people. He should always consult the dictionary when he is unsure about the meaning of a word. Sometimes words have acquired meanings different from the dictionary definition. These words should be used with caution.

Words have a social standing. Some words have snob appeal to some people and other words may seem cheap and vulgar to other people. The manager must consider the level of understanding of the group when he selects words.

Phrases that are familiar to the engineers in the organization may be misunderstood by others. Therefore, the manager must choose words that they can understand and meanings that will convey his ideas. He must use phrases that make his ideas easy for them to grasp. He must be concise, make sure each phrase is necessary for each part of the presentation, and remove excess verbiage. He can add forcefulness to his presentation by using strong opening and closing phrases. He should always use positive phrases rather than negative ones. In conveying his ideas, he should use action verbs such as "will" or "should."

The Presentation

The engineer manager must never offend the members of his group by attacking their self-esteem. First he begins by emphasizing thoughts and ideas on which there is agreement. Through this technique he tries to get the group to think positively. He may have to divide his idea into various segments. He then presents these segments, starting with the ones with which they will agree. This puts the group in a positive mood so that it will better receive the more controversial subjects later.

It is difficult to tell people they are wrong and have them accept it. Frequently, the manager can use an indirect approach by giving an example of undesirable behavior that has gone on or is going on in another department. This gives his group the opportunity to draw their own conclusions as to the performance desired by their manager. Conclusions thus drawn will greatly aid the manager in achieving his objective of behavioral change.

The engineer manager should have good posture. He should not lean or sit on the desk or put his feet on the chair beside the speaker's stand. When he is waiting his turn to speak, he should sit up straight and quietly in his chair. His facial expressions should convey a pleasant message to the group. He should develop the habit of always looking at his audience when he talks and he should smile often. A smile can never be overused, it never wears out, and it seems to grow and become more effective with use.

The engineer manager's actions should convey confidence and assurance, not hesitancy or uncertainty. Body and hand motions help the engineer manager communicate ideas to the group. Therefore, he should not put his hands in his pockets or wave them aimlessly. When he is not using his hands, he should let them hang relaxed by his sides.

The manager must be polite and courteous to his group. If there are any new members, he should introduce them first. In this way he helps integrate the group and creates a friendly atmosphere. He makes sure the shy and hesitant ones feel wanted and a part of the discussion.

The engineer manager must emphasize certain words or phrases in his presentation. These can be emphasized either by changing volume or by pausing. At times reduced volume is effective because it causes the group to listen more closely. He must give emphasis to timing because timing gives the group an opportunity to absorb what was said and prepare for the next statement. To add variety, he can speed up the utterance of less important words, thereby generating a feeling of importance and excitement. By cultivating such variety in his voice, the manager avoids a monotonous speech.

In addition, the manager should make use of the pitch of his voice. He may raise his voice to a higher pitch in order to give a feeling of excitement and lower his voice in order to give a feeling of confidentiality. Questions to the audience should always be asked with the voice rising to a higher pitch. Statements normally end with a lower pitch.

One of the greatest improvements that the engineer manager can make is to ensure that all words are understood by all members of the group. This is developed through practicing good diction. He must pay particular attention to the endings of all words, particularly the "ings" and "eds," and avoid slurring words together or blurring syllables. Good diction becomes a habit through practice. It will pay off by making a favorable impression on the group.

To be effective, the manager must maintain constant eye contact with individuals in the group so that he is the center of the group's attention. He captures the group's attention and holds it by looking them in the eye, talking to them, smiling with them. He avoids looking exclusively at one individual but allows his eyes to roam the room, from side to side and from front to back. He does not talk to the ceilings, floor, or windows.

The manager should avoid nervous gestures or other movements that will distract his audience. He should seek the criticism of his colleagues to flush out his distracting mannerisms. There are several things he can do, such as exercises to relax his throat and jaw muscles so that he can produce clear, full words. He can practice breath control, especially when he feels he is becoming tense. Good clear word production comes about through constant practice. There is only one rule for voice and word improvement—PRACTICE.

It is important that visual aids be used effectively. Some suggestions for effective use are listed below:

1. Keep aids in the correct order so that they can be used without disruption.
2. Avoid looking at the aid while speaking.
3. Allow the group to absorb the contents before starting to talk about it.
4. Do not stand in front of the aid.
5. Make sure the room is of sufficient size to use aids properly.
6. Do not make severe light changes; use a dimmer switch if possible.
7. Use only one thought with each aid.
8. Use different colors for illustrating different subjects.
9. Know how to operate the equipment or have a well-trained operator.
10. Use chalk properly, i.e., make horizontal and vertical strokes and make wide lines.

Group Feedback

For an effective presentation, the engineer manager should survey members of the group prior to speaking to determine their mood. During his presentation he should constantly observe their reactions. He should cultivate their interest as his presentation progresses and use key words and phrases. He should attempt to determine the opinions of the individuals in the group as the presentation progresses. In formal presentations he may obtain this only during the question-and-answer period. However, he can anticipate this response through feedback from the group by watching:

1. Eye movements
2. Body movements
3. Position of hands and arms
4. Position of head

All of these signs indicate how well the group is receiving his presentation. For example, when members of the group are constantly shifting positions in their chairs, their arms are folded, and their eyes are not on the speaker but on other objects, it is an indication that his presentation is not being received and accepted by the group.

Whenever possible, the speaker encourages members of the group to participate, thereby securing their ideas and attitudes. This gives him feedback and allows the group members to sound off. Also, it gives him an opportunity to adjust his presentation.

No one likes to be told what to do; people like to believe they are doing it on their own. Therefore, the manager should make suggestions but let members of the group develop their own solutions. He should encourage each member to contribute his own ideas, solutions, and plan of action. This makes the group much more willing to accept the final solution.

The effective manager knows that he cannot expect to receive full acceptance of his ideas. Therefore, he is always willing to compromise, especially on minor points. He must remember that there are many ways to accomplish the same objective; therefore, if one idea is more acceptable to the group than other ideas and will still lead to the same conclusion, by accepting the idea, the engineer manager has achieved his goal. Also, the group believes that it has determined how the goal will be achieved and that it was a very important part of the decision-making process.

Finally, the engineer manager should always attempt to achieve some kind of commitment at the end of his presentation. If it is impossible to achieve a decision, he should develop what the next step will be to secure a decision or agreement on an idea.

BEING INFLUENTIAL AT MEETINGS

Every engineer manager will be involved in conferences and committee meetings regardless of his likes or dislikes for this activity. His attendance at meetings is the quickest and easiest way for him and his superiors to form a judgment on his ability to sway others to his views. If he performs well at these meetings, he creates a positive image that everyone notices. But if his performance is poor, his future progress may be affected.

For the engineer manager to be effective and influential in the meeting, he must be positive in thought and behavior. He must carefully consider all proposals that are made and remember that there are many alternatives. If he cannot accept a proposal, he is better off not to come out adamantly against it. He will be wise to suggest another approach. He should realize that even though his reasons against the proposal might be valid and expressed professionally, the group will tend to see him as a negative person. But by making alternative suggestions, he will appear to be a "thinker," one who has a broad view of the subject.

The engineer manager should remember that all members of a group are relatively insecure and any criticism by another member makes instant enemies. Therefore, before the engineer manager speaks, he should always ask why he should be giving this criticism? Is it fair? Is he speaking for someone else? What will be its effect on his position? Certainly, if it does not directly affect him, he should avoid it. If others are unwilling to speak up for themselves, he should not become a self-appointed spokesman for them. When the group reaches a decision and the chairman states that this is what will be done, then it is time to cease the arguments.

Before attending the meeting, the engineer manager should prepare himself for his part in the meeting. If he must make a presentation, he should realize this is an opportunity to look good and gain respect. He should realize that he will be in front of a group that will be critical of his every word or phrase. This requires top performance, the best techniques that he can use.

During the meeting the engineer manager must be professional, competent, and aggressive. Before each meeting, he should review the agenda and if he is making a proposal, he should visit with key people to line up their support. He should try to find out how key people feel about major points. If the seating arrangement is unstructured, he should try to sit with the key people, the ones who are leaders. This will aid his image.

The engineer manager must be alert to what is going on in the meeting. He must avoid getting involved in arguments between groups or individuals. He should realize there are no rewards for playing a conciliatory role. When he makes suggestions to the group, he must be aware of thieves who may quickly respond, "I agree that it is a good idea. I'll work on it and prepare a proposal for the group." At this time the engineer manager needs to take immediate action. He should thank the other person for his offer and indicate that he has already begun work on the proposal but will be happy to consult with him if he wishes. Frequently, it is better to submit new and novel ideas in writing.

When the engineer manager is attacked in a meeting, he must maintain a positive attitude and respond in a calm and quiet voice. He should not make a statement that the individual is wrong; instead, he should put it positively, "I understand, but there are many other facts that you may not understand. We can talk about these when there is more time." Now the person is on the defensive and the engineer manager has made him appear to be an irresponsible individual who did not get all the facts.

During the meeting the engineer manager is well advised to:

1. Be tolerant of differences of opinion and consider all sides to the question or problem.
2. Pay attention and always be courteous.
3. Comment only when he can make an authoritative, knowledgeable statement.
4. Find out in advance the purpose of the meeting; be informed on who will attend and their points of view.
5. Cooperate with the chairman.
6. Take notes during the conference, particularly of people's viewpoints, for later reference and for future presentations.
7. Think before speaking; be brief.

Performance in meetings can make the engineer manager look either good or bad. It is important for his future that he performs well.

People react favorably or unfavorably to the characteristics of others. Some of the characteristics an engineer manager should have if he is going to be effective in influencing his people are listed below:

1. Talks about "us" instead of "I" or "my people"
2. Tells people how he can help them
3. Keeps his promises
4. Does not misrepresent
5. Knows his business
6. Helps people and trains his subordinates
7. Talks about specific problems
8. Is courteous
9. Knows people by their full names
10. Does not carry tales
11. Criticizes people in private and compliments them in front of others

SUMMARY

An important area of communication for the engineer manager is making presentations to groups. These groups may be made up of either superiors or subordinates. In making such presentations, it is important that he have a specific goal or objective. It is preferable if there is only one goal for each presentation. He will vary his presentation somewhat, depending on whether it is made to superiors or to subordinates. The variation is primarily in the choice of words and the manner of presentation. It is important that the presentation be couched in such a manner that it is compatible with the background and receptivity of the audience.

It is important to have a strategy associated with the method of presentation. This will alter not only the content of the presentation but also the order and manner in which it is presented. Second only in importance to this is a thorough understanding of the nature of the audience. The more the engineer manager is aware of their characteristics, the better he can make his presentation so that it will be well received.

He should make liberal use of visual aids because they greatly help to reinforce his oral presentation. When finally making the presentation, he must be aware of any irritating mannerisms that may distract his audience. It is important that he avoid and eliminate them by seeking the criticism of his peers or through a study of TV tapes.

Another area in which a manager performs is in meetings. It is important that he recognize how to use a meeting effectively so that he will present himself in the best light.

Important Terms

Presentation: A formal method of bringing proposals and ideas to the attention of superiors and subordinates.

Strategy: Developing the presentation to appeal to individual personal interests or the perceived group interests.

Styles of Behavior: These may be defined as detailers, affecters, strangers, and pushers.

Visual Aids: Supporting graphs, slides, or films used to emphasize and reinforce portions of the presentation.

For Discussion

1. Name several characteristics of the engineer manager that affect his degree of influence on groups.
2. What must the manager do before he can effectively present his ideas and influence others?
3. Name some factors which frequently cause the engineer manager to be unsuccessful in selling ideas to his subordinates.
4. Why should the manager have a thorough knowledge of his goals before presenting them to a group?
5. Is it necessary to have a knowledge of alternatives before making a presentation?
6. What role does strategy play in making presentations?
7. If the manager has a knowledge of the common interests of the group, how would he use it throughout his talk?
8. Name and briefly describe four different styles of behavior.
9. Name five things that the engineer manager should do in preparing for a presentation.
10. Name some ways that he can capture the attention of the group and hold it throughout his presentation.
11. What role does the visual aid play? Name some ways it could be used.
12. Name some things an engineer manager should be doing while he is making his presentation.
13. How can he determine the effect of his talk on the group?
14. Of what importance is the manager's performance at a meeting? How can he make this important to him?
15. Name five things he might do during the course of a meeting to strengthen his position.

Case 12–1

The Union Case

Sandy Miller was plant manager of the Metal Products Company plant in Aliquippa. This was a nonunion plant located in the heart of the unionized steel industry. For years Metal Products Company had tried to pay slightly higher wages than the surrounding steel plants and votes for unionization had been defeated.

Recently a new vote was called for, and this time the union was voted in. This was despite the fact that a wage increase exceeding the area average increase had just been granted.

Sandy was shocked. She wondered what had gone wrong. She was afraid she would be blamed since she had only been in this position for a couple of years.

She reviewed recent happenings and looked for any events that might have contributed to the change. She thought she had tried to listen to all complaints about working conditions. She had installed a suggestion system, but she had to admit she had not made any awards. The suggestions hadn't seemed very good and she hadn't bothered to set up a Suggestion Committee to handle awards.

Another strange thing was that the company had just announced a multi-million dollar investment in some new facilities that would bring about major changes in the way the plant operated. These changes should reduce costs and increase profits. Why wouldn't this make the employees happier?

    1. What went wrong at the Aliquippa plant?
    2. Why the sudden change of mind on the part of the employees?
    3. How should the new facilities have been presented?
    4. Why didn't the suggestion system work?

Case 12–2

The Delinquent Data

Paul was responsible for collecting certain efficiency and cost information from various plants operated by Metal Products Company. He worked for Rex Smith, Manager, R&D, and the data were useful in determining the need for efficiency studies, for identifying equipment that might no longer be performing efficiently, and in finding areas needing research work. Paul was to prepare quarterly reports for management analyzing the data.

His data came from 15 different plants located throughout the United States and Canada. He had no line authority over the plant manager and his ability to prepare and distribute good reports depended on this computation. Also, the quality of his reports depended on the quality of the data furnished. At times he felt they sent him "graphite" data rather than the correct data.

The situation seems to be deteriorating and he is finding it increasingly more difficult to get the data he needs to do his job.

1. What has brought about this situation?
2. What can be done to correct it?

For Further Reading

DIVITE, SAL F., "Selling R & D to the Government," *Harvard Business Review* (September–October 1965), 62–75.

HARRISON, ALBERT A., *Individuals and Groups: Understanding Social Behavior*. Monterey, Calif.: Brooks-Cole, 1976.

MURDICK, R. G., "How to Evaluate Engineering Proposals," *Machine Design* (September 28, 1961), 116–20.

"Selling to NASA," *Aviation Week and Space Technology,* (mid-December 1960), 36–37.

SINGER, JOHN, "Seeing Eye to Eye: Practical Problems of Perception," *Personnel Journal* (October 1974), 749.

TIME MANAGEMENT FOR THE ENGINEER MANAGER

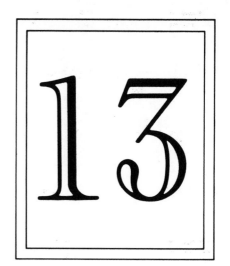

The engineer manager is under constant time pressure. This requires that he utilize his time effectively. This is especially important since time management affects the productivity of the engineers in his group.

LEARNING
OBJECTIVES

- ☐ Understand time as a resource.
- ☐ Gain insight into the engineer manager's work.
- ☐ Learn to identify and cope with time wasters.
- ☐ Appreciate the importance of using time effectively.
- ☐ Make and analyze time logs.

EXECUTIVE COMMENT

CARL J. WEIS
President
Bank Building Corporation

Carl J. Weis joined Bank Building Corporation as a buyer in 1949. He quickly moved up the organizational ladder, holding successive positions as project manager, manager of construction department, general construction manager, and vice president in 1968. Two years later he became group vice president, then senior vice president, and in 1976 president.

Time Management for the Engineer Manager

One of the major problems facing any engineer manager today, and in the future as well, is the "managing" of his time.

One major user of time today is "a meeting." "I'm sorry, Mr. Jones is in a meeting. Mr. Jones will return your call after his meeting." Business has always required and will continue to require the interface of people for accomplishment. If an organization is troubled, let me wager one of its major problems is poor communication. Communicating takes time if the understanding is to be complete. A major portion of one's time can be managed by the ability to communicate clearly, concisely, and with an understanding of the task to be accomplished. So, "a meeting" is not the culprit that devours time. It's the lack of ability to manage a meeting.

There are many helpful techniques that provide a path or direction to better time management. A few follow:

1. Creative time which covers anything you do that relates to planning your work, organizing it, and evaluating your own performance. It is the thinking part of the job rather than the doing part. You've heard the expression, "He's well organized, isn't he?"

2. Preparatory time which is the setup phase of work. Sometimes an entire effort can be lost through failure to spend a few extra minutes in getting organized. Don't waste your time or that of others by failing to set objectives for your meetings. Robbing others of time by your failure to be prepared makes you the "thief of the century" because of the precious nature of time. Your time and your talents are the principal talents you can contribute.

3. Productive time is premium time. Productive activities and the time you spend on them are the hard core of your work. They constitute the things which earn your living—the reasons for your employment.

4. Overhead time and overhead activities relate to your work as a whole as well as to your position in the organization and community. They include public relations, personnel relations, general correspondence and reports, and office paperwork.

Time is available equally to everyone. What matters is how well you use your time for the greatest return, whatever your objectives might be. In every walk of life the difference between the successful and the unsuccessful, in large measure, is how well they organize their work and time schedules.

Delegate all matters that do not require personal attention to responsible personnel. Use every resource available to multiply the impact of your best efforts. Effectiveness in delegation does not mean that you relieve yourself of all your tasks or that you rely entirely on assistants to decide what you should do. It is a reciprocal process of your telling them and their advising you. Out of this interaction comes a constructive balance.

When the engineer manager worked as an engineer, he was directed what to do and in many cases how to do it. In fact, when to do things was not a decision that he had to be concerned with, and how he spent his time was fairly well determined by his superior. But as an engineer manager, he is required to be largely a self-manager in the allocation of his time; therefore, it is essential that he and his staff be well organized. This focuses on how carefully the engineer manager uses the most scarce resource, "time," and on how he manages paperwork. The crux of efficiency is self-management of time. To the engineer manager, the managerial function increasingly becomes a greater and greater part of the company's total labor cost. Manufacturing activities account for approximately 25% of all U.S. employment, according to the Bureau of Labor Statistics, but by the end of the century manufacturing activity will account for only 5%. In the past, time studies concentrated on the manufacturing level, where the problem was easier to identify and the solution was easier to implement. In most organizations the managerial staff is very unproductive; they take many breaks, talk too much on telephones and in hallways, and read too much unrelated material on company time. Unfortunately, most engineer managers set profit goals, but few set time productivity goals so that they use their time effectively. The future of any organization or group is determined by the time productivity of each individual person.

The engineer manager must make sure that everything he does does not "just happen." Efficient engineer managers consider their talents and time as real assets, the same as other assets. They invest their time and talents by using the same criteria they use to invest the funds of the organization. The engineer managers who use time efficiently have learned a wide range of techniques for doing so and those who do not have them can learn them.

TIME AS A RESOURCE

Engineer managers must learn to think of their time and their subordinates' time as a very valuable resource, for "time is money." The only time they are making money is when they are working. Talking to friends, waiting, and traveling are things that *cost* money. Time is a resource that is unique in that it cannot be accumulated or stored like money. We are forced to use it whether we can use it effectively or not. It cannot be turned on and off like equipment or used by someone else. But, like other resources, it may be effectively or ineffectively managed and it can be controlled.

The engineer manager must allocate his time by following the same criteria he uses in making other engineering management decisions. In fact, he must allocate use of his time by following the same criteria used in investing for return on capital. Some activities will provide a greater return than others. Therefore, he must use his time on activities that provide the highest return. He cannot leave the disposal of his time unplanned or controlled by someone else; he must become the manager of his time.

Once the engineer manager understands that time is a resource, that time moves relentlessly on, and that it passes at a predetermined rate no matter what he does, he will realize how important time is to his work and success. Then he will be able to learn to manage his time effectively. This effectiveness can be learned through the study of time management concepts. Even though he feels that his problems are unique, they are in fact similar to the problems faced by all other managers.

When he starts to examine his time resource, he identifies factors that are external causes of his inefficient use of time. Normally, he will consider other people and conditions outside himself as the basic causes of his misfortunes. However, a new source is invariably identified: self-generated factors. It takes a painful reassessment and a willingness to be self-critical to recognize his own ineffectiveness. This step is necessary before he can learn to manage his resource of time.

COUNTERPRODUCTIVE WORK

Attempts to increase time productivity of the engineer manager and his staff often involve activities that are counterproductive and do not attack the basic cause of the problem. The purchase of the most expensive office equipment or a new computer system may not increase productivity at all; instead, it adds more work. Another counterproductive practice is to reduce staff by a given percentage or to institute a hiring freeze. These are ineffective ways to gain efficiency, because the units that are inefficient generally stay the same and the units that are efficient are greatly affected.

The engineer manager must identify the work to be done before he can measure the output of his people. He must review various operations to determine which ones are being performed effectively and which ones are not. Such surveys also establish job descriptions and set time standards for individual jobs. Increasing productivity is the major prerequisite to improving output, and it may be aided by recognizing good workers through promotions, bonuses, and increased salary. The poor performers may be retrained, motivated, or, if necessary, dismissed.

Significant improvements in productivity can be accomplished by creating an environment in which workers can identify problems and make suggestions for improvement. Suggestions may include the use of new or revised forms, ways to reduce the cost of phone calls, or ways to reduce the costs of reproduction. Productivity achievements should be shared with other groups in the organization, thereby bringing additional benefits to the corporation. Each group must know not only where they have efficient and inefficient operations, but also how they fit into the total corporate picture.

CHARACTERISTICS OF ENGINEER MANAGER WORK

To make effective use of time, the engineer manager must understand the nature of his work and compare it with that of others. In studying the work characteristics of management, it is interesting to note that there are remarkable similarities for all levels of people involved in managerial work, such as foremen, supervisors, and middle and senior management.[1] The manager's pace of activity will be rather constant throughout the day; handling pressing problems quickly, being constantly interrupted, and changing priorities. In fact, even after hours he is unable to escape his problems as he searches continually for new ideas and solutions. He is directly responsible for the success of his organization; therefore, his job is never finished. There are really no tangible mileposts; he is never sure he is successful, never sure new problems will not develop tomorrow. He can never forget his work even briefly, because there is always something else he should do. He always carries the thought that he can and should do more.

Even with the best-planned working day, the engineer manager must cope with a great variety of activities. Unfortunately, very important activities are fre-

[1]Henry Mintzberg, *The Nature of Managerial Work* (New York: Harper & Row, 1973), p. 24.

quently unpredicted, and trivial problems take a great deal of the engineer manager's valuable time. Hence, he must be prepared to cope better with these trivial problems. The manager gravitates toward an active work role, that is, activities that keep him current and that are specific, well-defined, and nonroutine. For example, reading mail is generally a burden because he finds that the information is old, it lacks immediate feedback, or only a small portion of it tends to be specific and of immediate use to him.

The engineer manager has a very strong attraction to oral media; therefore, a considerable portion of his time is allocated to oral communication. This amount of time may range from 60% to 80%. Figure 13-1 shows a typical distribution of time and activities for chief executives. These can be further compared to Table 13-1. Differences will depend on the position within the organization and the responsibilities of the position. Attention must be paid to activities that demand the major portion of time, e.g., meetings. These relationships must be understood in effective time management.

Most engineer managers do not like to write memos; they prefer face-to-face contact. The reasons are obvious: Writing is time-consuming and moves slowly, much time elapses before there is feedback, and many problems require instant information and decisions.

The engineer manager's work can be characterized by the following conditions:

1. Free time is unavailable; his mind is often on work when he leaves the office or goes home.
2. Most activities are fragmented and of brief duration; interruptions are common.
3. Most activities are current, specific, well-defined, and nonroutine.
4. Oral contacts are his main means of communication.
5. Telephone, unscheduled meetings, and formal meetings consume most of his time.
6. He has free interaction with a wide variety of people, e.g., subordinates.
7. He believes he has little control over what he does.

The engineer manager's work involves a wide variety of duties that give him an opportunity to (1) exert control over many initial commitments which lock him into a set of ongoing activities and (2) by exercising his leadership, take advantage of his managerial role. The crux of the problem in time mangement is what does the engineer manager do. His role provides the starting point for this analysis. The most common working roles of the engineer manager are listed below:

1. *Motivator.* He must provide the organization with energy and vision, reassurance and encouragement to employees, goodwill to his subordinates, and a productive work environment.

Distribution of Hours

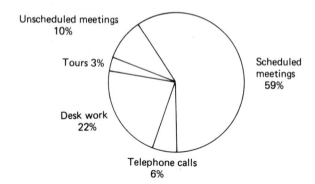

Distribution of Number of Activities

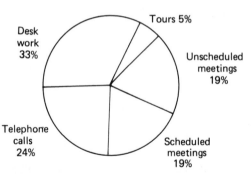

FIG. 13-1. Distribution of Time and Activities by Media (Based on Five Weeks of Observation of Chief Executives' Work). [Source: Henry Mintzberg, *The Nature of Managerial Work.* Copyright © 1973 by Henry Mintzberg. Reprinted by permission of Harper and Row, Publishers, Inc.]

2. *Obtainer of information.* He is always seeking information on the operation, always looking for operations that are going wrong and problems that need attention. He maintains a certain degree of alertness to actions in the organization.

3. *Representative.* Because of his position, he is obliged to perform many duties, some insignificant and others very important. Most of these duties are necessary, but they are not the center of his job. Many are considered a social necessity, for example, his appearance at certain events to add dignity and status.

4. *Organizer.* He must organize and lead the group through formal authority. The engineer manager has power; how well he organizes and leads determines how much of his power will be utilized.

TABLE 13-1

**Percentage of Engineer Managers Allocating Varying
Amounts of Time to Activities at Work**

Activity	Percentage Spending Less Than 30 Minutes per Day	Percentage Spending 30 Minutes to 1 Hour per Day	Percentage Spending 1 to 2 Hours per Day	Percentage Spending More Than 2 Hours per Day
Meetings	50	29	13	8
Telephone	32	36	28	4
Paperwork	8	16	32	44
Correspondence	20	40	28	12
Planning	17	38	13	32
Interruptions	26	48	26	0
Visitors	95	5	0	0
Social Conversation	64	36	0	0

Source: Michael LeBoeuf, "Managing Time Means Managing Yourself," *Business Horizons* (February 1980), 43.

5. *Maintainer of relationships.* Many relationships are maintained with his colleagues and with others at similar levels. He maintains a predictable, reciprocal system of relationships.

6. *Staffer.* Hiring, training, judging, promoting, providing remuneration, and dismissing are all part of the engineer manager's role.

7. *Observer.* He must be alert to changes, problems, and opportunities in the operation.

8. *Public servant.* He joins external boards, performs public service work, and attends conferences and social and political events.

9. *Analyzer.* He analyzes various issues, reports, and statements relating to both internal and external operations and events.

10. *Forecaster.* He uses a number of methods to better understand the trends in the industry and economy.

11. *Delegator of authority.* He assigns responsibilities and delegates the required authority for accomplishing the tasks.

12. *Instructor.* He gives subordinates information on how to proceed and trains them in decision making.

13. *Public relations specialist.* Various groups must be kept informed, such as boards of directors, suppliers, trade organizations, government agencies, customers, and the press. The general public must be kept informed of plans, policies, and results.

14. *Decision maker.* He solves problems, makes changes, searches for opportunities, and takes action.

15. *Supervisor.* He directs operations involving design, procurements, construction, and operation of facilities.

16. *Negotiator.* He may negotiate with unions on contracts or with his engineers on working conditions.

17. *Planner.* He develops and implements plans to achieve established goals.

IDENTIFYING TIME WASTERS   The engineer manager who does not practice time management finds that his time belongs to everybody else but himself. By examining his role and the characteristics of the engineer manager's work, he can identify various time wasters. Some common ones that he will immediately recognize are:

1. Unscheduled and scheduled meetings
2. Lack of adequate planning
3. Poor use of telephone
4. Poor delegation
5. Too much socializing
6. Ineffective communication
7. Assuming unnecessary public relations responsibilities
8. Lack of properly trained personnel
9. Lack of goals and objectives
10. Poorly organized supervision
11. Paperwork
12. Reading mail

Actually, the list is endless. The engineer manager must pinpoint those time wasters over which he has direct control. But just as important, he must understand some of the common myths about time management.

All new engineer managers constantly hear that to get ahead takes "hard work," which implies long hours, and that as he is promoted, he will assume greater responsibilities and have a greater span of management. However, each promotion brings more authority, which allows more delegation to get the work done through others. When the engineer manager puts in long hours at his office, does not take regular vacations, and spends little time with his family, he also fails to have time available to develop the creative aspects of engineering management that are important to his success.

It has been found that productivity declines rapidly when one works more than 8 hours and when a meeting lasts longer than 2 hours. When the manager works each evening, he soon acquires the habit of delaying important work till then. This means that what should be done in an 8-hour day is often stretched into a 10- or 12-hour day. He sacrifices his leisure time, family, and personal health and soon his job performance is lowered. There is no direct relationship between hard mental work and positive accomplishment. In fact, an engineer manager who is unproductive may well attempt to offset this by appearing to work hard or maintaining long hours. Results are seldom in direct relationship to the amount of time spent.

When the engineer manager is very active, it is not uncommon for performance to be inversely proportional to the level of activity. If he is insecure and lacks direction, objectives, and self-confidence, he tries to make up for it by vigorous activity. Consequently, he must exercise control to distinguish between activity that gets results and misdirected activity. He seems to work long hours by choice because of fear and his inability to use time effectively. Part of the drive seems to be the result of his constant uncertainty about what to do with leisure time.

It is not enough that the engineer manager knows his time wasters; he must find out the reasons for them. Some common time wasters are shown in Figure 13–2. Time wasters by definition show that something more important is not getting done.

COPING WITH TIME WASTERS

After identifying the time wasters, the engineer manager must now direct attention to their basic causes. These habits are a bane, not a blessing, to the engineer manager, the organization, and subordinates. An evaluation of the engineer manager's work habits and time management skills inventory (Tables 13–2 and 13–3) indicate where corrective action needs to be taken. Some definite actions the engineer manager can take are listed below:

1. Use a time inventory and prepare a daily time log.

2. Determine the tasks that are meaningful to the accomplishment of the organization's ultimate long-range goals or objectives.

3. Divert all tasks to areas of responsibility that are directly assigned to designated individuals.

4. Determine areas or functions most important to the engineer manager's job; these should include the following:

    A. Planning for the future of his group

    B. Coordinating the various areas and functions

    C. Participating in activities as a representative of the organization

5. Delegate all repetitive tasks. When the subordinate says, "We have a problem," the engineer manager must consider carefully whether or not to get involved. The manager may not know enough to make the on-the-spot decision expected of him. If he responds "Let me think about this problem," he has taken the problem from the subordinate, and the delegation process is broken.

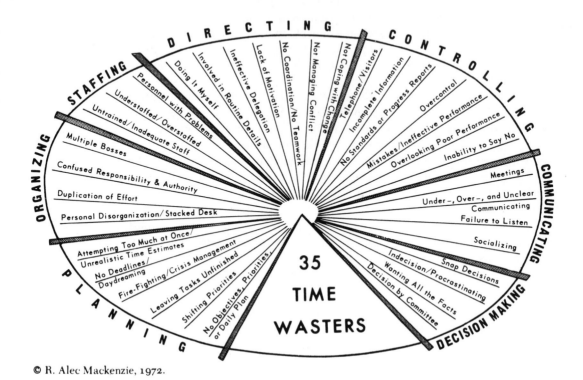

© R. Alec Mackenzie, 1972.

FIG. 13-2. Common Time Wasters According to Managerial Functions.
[Source: R. Alec Mackenzie, *The Time Trap* (New York: AMACOM, 1972),
p. 86.]

6. Do all similar tasks at one time and at the most opportune time. For example, he should make all telephone calls at one time. This allows him to prepare for the calls, identify their purpose, and make them briefer, since he will be aware of how many there are to be completed in the allotted time.

7. Identify the time of day when he does his clearest thinking and identify the best time for dealing with specific activities. Develop priorities according to the best time for dealing with them.

8. Practice self-discipline, follow time management rules, use strict deadlines, and exercise tight control.

TABLE 13–2

Evaluation of Work Habits

Answer each question either "yes" or "no."

1. Are you a self-appointed witness of all the organization's activities?

2. Do you assume everyone's responsibilities?

3. Do you always have a messy desk?

4. Is there a disarray of important papers and memos?

5. Do you have lunch at your desk, not take breaks, or not leave the office because things will go wrong?

6. Do you take home a briefcase full of work every evening?

7. Are you the last to leave the office?

8. Have you missed taking your vacation?

9. Do you take only emergency trips?

10. Are you unable to meet deadlines?

11. Do you not have time for your family?

12. Are you always volunteering to do a job?

If more than three questions were answered "yes," you have very poor work habits and your time is being used ineffectively.

TABLE 13–3

Time Management Skills Inventory

The ten questions below will help you rate your basic time management skills. Answer each question either "yes" or "no."

1. Do you always prepare a daily written "To Do" list?

2. Do you always prioritize your "To Do" list?

3. Do you always meet deadlines without having to rush at the last minute?

4. Are you on top of your paperwork?

5. Do you ever let interruptions sidetrack you from your "A" task?

6. Do you always tackle hard or unpleasant tasks without procrastinating?

7. Do you always accomplish what you want to get done during the day?

8. Do you know the difference between efficient and effective use of your time?

9. Do you spend too much time on trivial matters?

10. Do you feel you have enough time for planning?

By using a time log, the engineer manager will find a number of unsuspected enemies usurping his time. He will also find that his time is generally wasted in the same way every day. Coping with time wasters is not the easiest job the engineer manager faces; in fact, most managers fail to pinpoint and recognize them as difficult problems. But once the manager has identified these time wasters, he will find that there are many areas in which he can improve his effectiveness and cope with the major time wasters. Some of the things he can do are listed below:

1. *Correct the work environment.* When giving serious consideration to better managing his time, the manager must examine the physical layout of facilities. The physical location of his office compared to that of his secretary will determine the number of visitors screened. Certain subordinates may be required to be near the engineer manager because he needs close contact with them for his work. The location of subordinates is a very real factor in determining frequency of communication. Being in a different part of the building or on a different floor is a significant contact barrier.

2. *Reduce distractions.* The proximity of desks to each other causes unavoidable noise of voices and typewriters. This noise can be reduced by soundproofing. Angular partitions in cheerful colors, carefully selected and blended for a pleasing appearance, make concentration easier and improve morale.

3. *Improve equipment.* Uncomfortable seating and dim or flickering lights can lead to physical fatigue and reduce productivity. Replacing a secretarial chair, table, or files that are not properly designed can be the most valuable investment an engineer manager can make to increase productivity. Even a shoulder support for a telephone can become an extremely useful tool.

4. *Do not use a desk.* Not having a desk in an office should be considered, for the desk is a formidable barrier to communication. Chairs arranged around a coffee table inspire communication; a stand-up desk is an effective alternative. Standing up and walking around during a conversation give a different perspective. Not having a desk makes a first-line supervisor get out on the floor where the action is. It prevents him from staying in his office and waiting for problems to come to him.

5. *Clean off desk.* Many an engineer manager has a desk stacked with things he does not want to forget. This works for a while, but soon the stacks grow so large that he is unable to find anything. Every time he looks at the stacks, his train of thought is broken. Since he can only do one thing at a time, he should clean the desk off so that only one problem is worked on at once. All other problems should be placed in the file. He should finish one problem before starting work on another. He should make it a responsibility of the secretary to have the material needed for each problem on his desk when he asks for it and have the next problem ready according to priorities.

6. *Sort mail.* The art of sorting out important mail is one of the most critical skills. The engineer manager finds himself reading junk mail in order to determine whether or not he ought to be reading it at all. He also has a tendency to overfile, which is costly both in time and space. To overcome this, the engineer manager should skim the mail and lay aside all important pieces for further re-reading and studying. The rest should go in the wastebasket. Since most of the mail should probably go into the wastebasket, the wastebasket should be large enough and conveniently located. The decision about which mail should come to the engineer manager's desk can be made by his secretary.

7. *Develop a filing system.* All too often the office's filing is left to the engineer manager's secretary. Consequently, very often the filing system is a source of constant irritation and it is monumental time waster in terms of retrieving information, except for the secretary who developed the system. To make files work efficiently, the engineer manager must have a uniform approach so that everyone can use the files, write instructions on the filing procedure and provide a diagram and guidesheet of the system, and develop a policy on the number of years documents are to be saved. Current files ought to include no more than one year's correspondence.

8. *Use memory joggers.* The engineer manager should set up a system for following up items or tasks on future dates. This is an invaluable timesaver and memory jogger.

9. *Maintain a pocket diary.* Every engineer manager has lost or misplaced information or has made a futile search for notations. This timewaster can be eliminated by keeping a pocket diary for taking notes. He should keep it with him at all times; he should even put it on the nightstand beside his bed so that he can jot down thoughts during the night, for many a time he won't be able to remember them the next morning.

10. *Develop short responses.* A few sentences are all that are required for most memos and letters if they are answered immediately.

PLANNING TIME

Being busy is simple for the engineer manager but being effective is difficult. Planning activities is a necessary requirement because he does not have time to do all things that his conscience or imagination tells him he needs to do, but he must decide what *to do* and what *not to do.* Actually, the proportion of "must-do" to "like-to-do" activities is relatively small. The solution is not working long hours. Instead, it is setting priorities. Otherwise, he will constantly put second things first by default. Then he will be in trouble.

The engineer manager cannot do all of the jobs that come to his attention. Therefore, he must decide what is important and do that well. Refusing to do unimportant tasks is a sign of success. This requires the engineer manager to stay out of day-to-day operations and plan for the future. He must have his subordinates do the more routine tasks of the organization. He has a very basic need to know, but he must protect himself from the mass of operating detail that a typical information system provides. He must learn to avoid information that (1) does not inform, (2) provides details on which no action can be taken, (3) does not provide all the necessary details concerning the situation or activity.

Setting priorities, even for the next day, is difficult for the engineer manager because he feels it limits his freedom. Besides, most good engineer managers pride themselves on being natural leaders who can make decisions by intuition instead of thoughtful evaluation. Setting priorities requires planning time in order to visualize not only the events that he wants to take place but the various alternatives for accomplishing them. This becomes important as work expands and grows more complex.

Yet barriers to setting priorities are numerous. The emphasis on day-to-day activities almost always pushes established priorities into the background. Consequently, coping with today's problems takes precedence over the established priorities. Uncertainty about the future is also a deterrent to the accomplishment of the priorities. The engineer manager feels more comfortable working within a structured situation in which factors are certain and predictable; therefore, he tends to avoid other situations that may be more important. Yet as he moves into higher organizational levels, setting priorities becomes more important and to a great extent determines his effectiveness. Although the following activities are necessary for the engineer manager to establish priorities, he resists them:

1. Taking the time
2. Giving the thought
3. Following the procedures
4. Giving the commitment

Setting priorities (Table 13–4) will at first seem time-consuming and he will wish he could forget it and get back to work. He will be discouraged when the time required seems out of proportion to any immediately identifiable benefits. In fact, the engineer manager really feels he does not have time to even think about setting priorities.

The urgency of the tasks at hand usually takes priority. Unfortunately, the more crises, the more issues facing him, the less he is likely to take time to set priorities. Yet, setting priorities is his only real hope. For every engineer manager who has succumbed to the pull of the day-to-day crises, there is another manager who has fought the battle through and has come to recognize that survival depends on setting effective priorities. After the results are clear, it will be evident that this is the most essential technique available to the engineer manager. Otherwise, he will be engulfed in paperwork and other unrelated activities.

TABLE 13–4

Essential Steps in Setting Priorities

1. When starting a new job, reexamine and analyze priorities.
2. Develop relevant information about existing conditions and the time frame.
3. Establish or reexamine goals and objectives.
4. Set priorities to ensure that:
   A. Adequate time is given to each priority item
   B. Crises do not determine the priority
   C. Adequate information is obtained
5. Implement priorities with the following in mind:
   A. Do not do the interesting jobs first.
   B. Do not do a task because of a telephone call.
   C. Do not do additional jobs in order to get them out of the way.
6. Review each day to determine if priority activities were accomplished. If they were not, determine why and reorganize so that they can be achieved.

Urgency always dominates the manager's time and influences his priorities, but the most urgent tasks are not always the most important. Those urgent matters, however, distort the engineer manager's priorities; they make minor projects take on major status under the guise of crises. As a result, the urgencies cause him to "manage by crisis." He must develop the ability to distinguish the important from the so-called urgent and not be tyrannized into departing from his order of priorities.

One of the major reasons for failure to plan is that today's crises are considered more important than the planned priorities. To the manager, the urgent tasks appear to require instant action, and these tasks seem irresistible at the moment, but if he does them, they will consume all of his time and energy. Rarely do these urgent tasks have to be done today; in the long run their deceptive prominence fades. When his priorities are pushed aside by these urgent activities, he becomes a slave to insignificant time-consuming activities. Consequently, these crises manage the organization rather than the engineer manager managing them.

Few engineer managers will quarrel with the importance of setting priorities for their activities. But most will admit that the pressure of events and crises conspires to prevent them from fulfilling their planned priorities. This weakness must be overcome if the engineer is to be successful as a manager. This is why faithfully learning and following time management concepts is necessary.

MAKING EFFICIENT USE OF TIME

We all have the same amount of time—24 hours every day. The difference between us comes in how effectively we use it. Unless the engineer manager manages his time effectively, no amount of ability, skill, experience, or knowledge will make him a success. He needs training in the skill of time utilization to avoid crises and to perform activities that benefit the organization. Ways the engineer manager can use his time more effectively are to:

1. ***Do important tasks first.*** First decide what is important. Generally, one does the things that he enjoys first and procrastinates on the tasks he dislikes. By preparing a "To Do" list each day and assigning priorities, and then doing the first priority first, the engineer manager begins to get control of his time (Figure 13–3). He must stick with the first priority item until it is finished. Large tasks can be subdivided before priorities are assigned.

Daily Tasks Log

Date _____

Priority	Tasks	Estimated Time
1		
2		
3		
4		
5		
6		
7		
8		
9		
10		
11		
12		
13		
14		
15		
16		

FIG. 13–3.   Daily Tasks Log

2. ***Delegate.*** After determining priorities and dividing the task into manageable sizes, he should examine tasks that can be delegated to subordinates. The engineer manager must break the do-it-himself habit and delegate whenever possible.

3. *Group similar tasks.* By grouping similar tasks, the engineer manager will be effective and will better utilize his resources and efforts. For example, seeing people all day creates unending interruptions. By grouping appointments all at one time during the day, he can be more effective in handling them. Soon people will develop the habit of seeing him only at this time, not whenever they prefer.

4. *Start early on important tasks.* The engineer manager should start the day with the first-priority tasks when his energy level is highest and he is most effective.

5. *Get organized.* He should clear the desk and set up a filing system. He should do only one task at a time. When his desk is piled with papers, the engineer manager wastes time looking for buried items.

6. *Reduce paperwork.* He should eliminate all unnecessary mail, throw out junk mail, cancel unused subscriptions, and have some of the mail go directly to subordinates. He should move paperwork along instead of letting it stack up on the desk.

7. *Control meetings.* Meetings disrupt work of both the engineer manager and his subordinates. Therefore, meetings should only be held when necessary, and they should follow an agenda. So that meetings will end on time, they should be scheduled near the lunch hour or near quitting time so that people will leave.

8. *Use waiting time.* He should plan tasks for idle periods. Waiting time and travel time can be converted into useful time.

9. *Do not try to do everything.* There is a limit to the things an engineer manager can do effectively. He must turn down unimportant requests. Otherwise, he will have little time for the important tasks.

The engineer manager's ability to manage his time effectively will determine his success.

MAKING A DAILY TIME LOG

The engineer manager is always surprised at how little time he has to accomplish what lies before him. Consequently, he must first begin by finding out where his time is being used. A time log (Figure 13–4) will give him an analysis of how his time was used and help convince him where he is wasting time every day. It will help him to plan his work periods, reduce interruptions, and remind him of the things he needs to do.

One can organize his tasks by priority and do them in that order and thus use his time more effectively, except that he has a boss who may destroy his best laid plans. So he must make an amendment to his work: "within the limits that my superiors impose . . ." The engineer manager is alone in judging his own performance. The task of self-evaluation is difficult, but a time log is an effective tool to use to make this evaluation. The purposes of a time log are to:

TIME ANALYSIS

Time	Priority Number	Activities	Comment
8:00			
8:30			
9:00			
9:30			
10:00			
10:30			
11:00			
11:30			
12:00			
12:30			
1:00			
1:30			
2:00			
2:30			
3:00			
3:30			
4:00			
4:30			
5:00			
5:30			
6:00			
6:30			
7:00			

FIG. 13–4. Time Analysis

1. Show how time was used
2. Allocate time
3. Put priorities on tasks
4. Evaluate effectiveness, that is, compare the time spent with the estimated time

Below are specific instructions for using a time log analysis:

1. At the end of each day list the tasks to be accomplished the next day and arrange them in order of priority (Figure 13–3). Next, estimate the time allocated to each task.

2. As the day progresses, record the results achieved for every 30-minute period (as in Figure 13–4). These results must be recorded as the day progresses, for memory is deceptive if recording is left till the end of the day. In Figure 13–4 the engineer manager has recorded the priority number and activity. If necessary, comments can be made on the tasks.

3. After the time log has been kept for at least a week, preferably two weeks, add the total hours spent and the total hours estimated for the time effectiveness evaluation (Figure 13–5). Then compare the estimated time with the time spent.

TIME EFFECTIVENESS EVALUATION

Priority Number	Estimated Time	Time Spent	% of Estimated Time to Time Spent

FIG. 13–5. Time Effectiveness Evaluation

4. On the basis of time effectiveness evaluation, determine the effectiveness of each task and the characteristics of the tasks.

5. The time log shows the engineer manager the areas of greatest effectiveness and the areas of least effectiveness. From these he can develop a strategy for improvement which should be implemented immediately.

The engineer manager will discover the following truths after using the time log system for two or more weeks:

1. A great deal of work is repetitive and can be delegated to others.
2. Many meetings and activities involve the same people on the same subjects and could be eliminated.
3. Very little time is available for thinking and planning.
4. The time log helps in scheduling and meeting deadlines.
5. The time log helps identify those who initiate meetings.
6. The time log helps determine what activities are unnecessary.
7. The time log shows that meetings consume more time than is desirable and will lead to reductions.

Since the engineer manager may have lived with many time-wasting habits for years, even if he succeeds in his efforts to change, he will find he has a tendency to gradually return to his old habits. Therefore, he must use the time log system periodically to check his effectiveness and correct his bad habits.

SAYING "NO" Learning to say "no" is a very difficult but common problem for most engineer managers. Although they complain about having more things to do than they have time, they still cannot say "no" to anyone, not even to their subordinates. The problem facing every engineer manager is how to say "no," but still make people feel motivated and want to try to do their best. To further complicate the situation, every project and priority has an advocate and is on somebody's "must" list. Below are some helpful suggestions the engineer manager can follow to deal with this situation:

1. Develop objectives with the group, and in turn, priorities. Then when someone comes with a project, the engineer manager can ask, "Is this our first priority?"

2. Do not agree to solve every problem for subordinates. Give them an opportunity first. The engineer manager will be surprised at what they can do.

3. When someone calls to see him, he should go to the caller's office because the visitor is the one who has control. It makes it easier for the manager to say no or leave on time. It also is good public relations and he will be surprised at what other things he will learn.

4. Make the person requesting help think and come up with alternatives. The engineer manager must be courteous at all times and praise the individual for thinking and making alternate suggestions.

Engineer managers must learn to think of their time and their subordinates' time as a valuable resource. The efficient use of time becomes an important objective for them.

To make effective use of time, the engineer manager must understand the nature of his work in comparison with that of others. It is generally true that there are great similarities in managerial work at all levels of management. The types of decisions they handle and the interruptions they receive are quite similar.

The engineer manager's work reflects a wide variety of duties, which not only give him the opportunity to exert control over many initial commitments but also to exercise leadership in their achievement. It is important that he should look at his role in these areas. He will find that his obligations are very diverse.

To improve time utilization, the manager should identify time wasters. Elimination of these will be helpful in improving his time utilization. There are a number of definite actions that he can take, including preparation of a time inventory and a daily time log to cope with these time wasters.

There are many areas in which the manager can improve his effectiveness, including correcting of the work environment, improving equipment, keeping a clean desk, developing a filing system, and using memory joggers.

Planning time is the essence of effective time management, just as it is in the effective management of all resources. The most important step is to list those tasks that must be done each day and assign each task a priority number. It is then important to complete the first priority item before moving on to the next one. The manager must avoid substituting a crisis of the day for one of his planned activities. If he allows himself to get into this habit, he will lose all control of his time.

It is important to recognize that we all have the same amount of time. The difference comes in how effectively we use it. A few things that will help the manager use his time more effectively are to do important tasks first, delegate, group similar tasks, start early on important tasks, get organized, reduce paperwork, control meetings, use waiting time, and not try to do everything.

Preparing a daily time log, making time analyses, and readjusting estimates are helpful ways to gain control of time.

SUMMARY

Important Terms

Counterproductive Work:	Efforts to improve productivity that may produce negative results.
Time as a Resource:	Time is similar to all other resources except that is possesses the unique characteristic that it cannot be accumulated or stored.

Time Log:	Device for recording and analyzing time utilization.
Time Wasters:	Distractions resulting from poor planning or outside interference that prevent the completion of the first priority items.

For Discussion

1. Discuss why time is an important resource.
2. Name some things that might be counterproductive in trying to improve time utilization.
3. Why is it important that the engineer manager recognize the character of his work?
4. Discuss the value of oral versus written communication for the manager.
5. Name and define ten roles that the engineer manager may play.
6. Name six time wasters that the manager encounters.
7. Discuss the effect of carrying work home at night.
8. Discuss six ways to cope with time wasters.
9. Name and discuss five areas of operation in which the manager can improve his effectiveness.
10. Why is planning of time utilization important?
11. What technique allows us to assign time?
12. Name five steps that are essential in setting priorities.
13. Identify five ways the manager can work to use his time more effectively.
14. Discuss the preparation, utilization, and analysis of a daily time log.
15. Discuss techniques for learning to say "no."

Case 13–1

Time Waste

Randy, one of Rex's lead engineers is a friendly, talkative individual. He loves to tell stories and he frequently has his team of engineers in stitches. This might not be too bad, but the commotion is such that other engineers tend to join in.

Randy further complicates the problem when he has to visit other engineers to secure information. Invariably, he tells a story and thus causes disruption in that group.

On the plus side of the ledger, Randy is a hard and effective worker. When his team is given a problem, they provide an excellent quality and quantity of work. Nevertheless, since Rex is facing a problem of several of his projects being behind, he finds Randy's gregariousness irritating.

Rex calls Randy into his office. "Randy, I'm well aware of the time you waste during the day telling stories to your team or anyone else who will listen. You and your team would fare much better in merit raises if I didn't feel you were disruptive to the department's operation."

"Well, Rex, I believe money is only good if you can enjoy it. Since we spend 25% to 35% of our life working, I figure we ought to enjoy that time. Life is too short not to."

However, Randy agrees to try to be less disruptive. He is for a few days but then reverts to his old pattern.

    1. Did Rex handle this problem professionally?
    2. Do you agree with Randy's philosophy?

Case 13-2

Time Management

Rex Smith was finding himself more and more bogged down in his new position as manager of R&D for Metal Products Company. He was responsible for all of the research and development activities, as well as, design, and installation of new facilities for the company. In addition, his staff prepared designs and specifications for the sales department as needed for engineered products. As a result of these varied responsibilities, many internal and external documents came to his desk.

Rex felt that he was becoming a bottleneck in the operation. He noted that they were beginning to miss some deadlines on projects because he was failing to find time to follow up on them. He was also discovering that the quality of their work in cost estimating and design did not seem to be up to earlier standards. He had begun to ask himself how much of this problem was his fault.

He knew that the continual backlog on his desk was keeping him from prompt analysis and approval or rejection of projects that came to him. He also knew that he was not getting enough time out of the office in order to keep his finger on the pulse of the operation and be alert to any bottlenecks along the way. To make matters worse, because of the many areas in which he was involved, phone calls and discussions with employees who dropped into the office always seemed to interrupt his studying reports. Even the employees were a problem because they seemed to drop in unannounced to ask questions or briefly report on their work progress. It seemed as though he was the focal point of everything that was happening in R&D and that nothing moved until he gave an O.K.

1. What problem does Rex face?
2. How should he set about getting his time under better control?

For Further Reading

BAKER, H. KENT, "Invest Time to Save Time," *Industrial Launderer,* (July 1977), 34–37.

———, "The Time Budget: A Personal Planning Tool," *Public Telecommunication Review* (May–June, 1977), 27–30.

BARRETT, F. D., "Everyman's Guide to Time Management," *The Business Quarterly* (Spring 1973), 72–78.

HEYWOOD, JAMES D. "Manage Your Time by Managing Your Activities," *Supervisory Management* (May 1974), 2–8.

"If You Don't Have Enough Time, Read This," *Changing Times* (March 1980), 50–51.

LAKEN, ALAN, *How to Get Control of Your Time and Your Life.* New York: Peter H. Wyden, Inc., 1973.

MACKENZIE, R. ALEC, *The Time Trap.* New York: AMACOM, 1972.

———, "Toward a Personalized Time Management Strategy," *Management Review* (February 1974), 10–15.

ONCKEN, WILLIAM, and DONALD L. WASS, "Management Time: Who's Got the Monkey?" *Harvard Business Review* (November–December 1974), 75–80.

SCHWARTZ, ELEANOR B., and R. ALEC MACKENZIE, "Time Management for Women," *Management Review* (September 1977), 19–25.

"Teaching Managers to Do More in Less Time," *Business Week* (March 3, 1975), 68–69.

WILKINSON, RODERICK, "Six Tips on Investing Your Time," *Supervisory Management* (September 1973), 30–33.

ENGINEERING MANAGEMENT TOMORROW

PART IV

LAW, ETHICS, AND THE ENGINEER MANAGER

The engineer manager practices his principles and concepts in an environment constrained by the law, morality, and the organization.

LEARNING
OBJECTIVES

☐ Learn the major sources of law which determine the legal environment.

☐ Understand the judicial system as it functions in this country.

☐ Learn the elements and critical stages in filing a lawsuit and in the conduct of a trial.

☐ Develop an appreciation of the significance of technical innovation to a company and the United States economy.

☐ Be able to describe the means of protecting technical innovations.

☐ Understand the role of licensing in managing technical innovations.

☐ Learn the many ways ecological laws affect management decisions.

☐ Recognize that philosophical ideas influence modern business practices.

☐ Understand why corporate social responsibility is of great interest.

EXECUTIVE COMMENT

ROBERT H. WIDMER
Vice President
Science and Engineering
General Dynamics Corporation

Robert H. Widmer is vice president of Science and Engineering for General Dynamics Corporation. A native of Hawthorne, New Jersey, he holds a B.S. in aeronautical engineering from Rensselaer Polytechnic Institute and an M.S. in the same field from California Institute of Technology. He joined Consolidated Aircraft Corporation, the predecessor of General Dynamics in 1939. He moved through a number of management positions to become chief engineer in 1959 and vice president of Research and Engineering in 1961 and to his present position of vice president of Science and Engineering in 1974.

Engineering Management Tomorrow

Engineering management can take on many forms and it means different things to people within or outside an engineering organization. I see it as both the stimulus and the binding force within engineering to accomplish its basic responsibilities for conceptual design, fabrication, and functional operation of products and systems of practical use to society. In the past the success of management was mainly judged by its products being "under budget cost," "ahead of schedule," and having "superior performance." In the 1980s engineering management will be judged by more complex criteria such as environmental considerations, energy efficiency, safety, ethics, and so forth. These and other new emerging criteria will be a real challenge for management to incorporate and keep in proper balance with the more understood requirements. Although today's engineering manager has available to him many, maybe too many, computerized management information tools, they by themselves will not eliminate the need for the basic personal characteristics of (1) leadership, (2) ability to make decisions, (3) technical respect of fellow employees, and (4) actions that track verbal and written policies. These characteristics together with organizational structures that are flexible and adaptive to the varying needs and demands of specific programs are the ingredients necessary for success in the coming decade and for restoring society's doubts on technology as the cornerstone solution for many of today's problems.

ENGINEERING MANAGEMENT TOMORROW

The engineer manager of the 1980s and beyond faces an environment in which he must operate that will be quite different from the environment of the past. Francis E. Reese stated it well when he said, "The complexities and difficulties that face not only today's but tomorrow's needs are almost frightening. Those in engineering management twenty years—or even ten years ago—probably share some of my feelings that we never realize how easy and simple it was. Yet, the demands put in front of us by the changing societal, economic, and political forces provide a challenge that is far more exciting and more involved than any of us could have envisioned twenty—or even ten—years ago."[1]

In the past decades the engineer applied his skills without serious concern for the effect of his creation on the environment. He was generally oblivious of government regulations and their effect on the success or failure of the venture. Not too many years ago if the corporation had the technological know-how and the finances and if its new product met a market demand, it could be assured of success. This is no longer true today.

Take as an example shale oil production. Many processes are available today which appear competitive with crude oil brought in from foreign sources. Financing is available to build the plants from private sources. But not one plant has been built because governmental restrictions through environmental regulations make it impossible to assure the investor that the plants will ever run. So the engineer today must be aware of societal and governmental actions that dictate concern for the environment and the world in which we live. He must design facilities based on known and new technologies so that they integrate with the economic and societal systems of our world. The economic impact of these actions on capital requirements and the operating costs are making the engineer a vital participant in the allocation of financial resources in industry.

The engineer is now a key contributor to decisions on how the world in which we live can be kept alive and well. The potential for engineering leadership and innovation has never been greater or the hazards of inept engineering so evident.

[1]Francis E. Reese, "Engineering Management and the Winds of Change Factor in 1977," *Professional Engineer* (February 2, 1977), 14–17.

Engineer's Sensitivity to Society's Needs

The engineer in the future must have a heightened sensitivity to society's needs. He cannot do an appropriate job without an intelligent and a forward-thinking perception of tomorrow's society. He must be as sensitive to these needs as the marketing specialist is to the product development man. His ability to handle formulas, facts, and scientific relationships must be matched by the ability to understand environmental and social needs and trends.

All of this underlines the importance of the engineering manager who must develop a team that can perform effectively in this new environment. He must add to his team others whose contributions are in skills and disciplines not traditionally considered part of the engineering team. Satisfactory performance will depend on the engineer manager's ability to integrate the efforts of his engineers with representatives of the financial, personnel, legal, and political fields. Each of these people will play a crucial role in the success or failure of his engineering efforts as viewed by society. The engineering team membership must be expanded to meet the larger responsibility that has been thrust on it.

Effect of the Economy

Because of the rapid acceleraion of material and construction costs, the economist and the financial analyst become important to the engineer. He must be aware of, and sensitive to, the economist's predictions of inflation. A decade or two ago the engineer paid little attention to these predictions. Today he cannot survive in his profession without taking them into consideration.

During the coming years the engineer manager must face the option of choosing from several equipment sources and he may be involved in the international scene and be concerned with methods of financing through the international bank. The feasibility of the project may alternately hinge on where to purchase the equipment, what currency to use to pay for it, and how to finance it.

Because society's wishes are expressed through governmental laws and regulations, the engineer manager finds himself with another member on his team—the governmental representative who must ensure compliance with environmental, health, and safety regulations. Unless disagreements with him on engineering design are appropriately resolved, the project will not go forward. Failure to develop feasible alternatives economically may kill the project.

Impact on Governmental Regulations

It is important that the engineer manager develop skill in working with governmental agencies. He should share an equal concern in the responsibility for tomorrow's world. Consequently, he should search for ways to meet regulatory demands as well as to meet his responsibilities to the shareholders of the firm. An important part of his work may be in developing facts that will provide compromise regulations that will meet the needs of both parties.

No less important than meeting environmental pressures is the need to meet the pressure for energy conservation. The escalating price of energy will make the need for conservation stay with us for the foreseeable future. The engineer must design with this fact in mind. He must be concerned with the economic balance between extra costs for facilities to meet conservation needs versus the money that can be saved through this economy.

To meet these ever-increasing and more complex needs, the engineer manager must concern himself with the source and quality of engineers. He should see that they are well trained to have this broader understanding of their role. Frequently he will have husband and wife working in the same corporation or possibly in the same department. He must develop the best plan to utilize their capabilities effectively. He must assure himself of the necessary backup personnel in order to assure continuity of engineering participation from design through the startup of a project.

Updating of Engineers

Over the years he must concern himself with retraining and updating the skills of his engineers. This will be necessary because the technology base will continue to escalate. It has been said that there is a ten-year half-life to current engineering knowledge in this increasingly complex world that requires environmental protection and energy conservation. People have a tendency to feel that technology is infallible and that it should give instant solutions to society's challenges. Society also expects individuals to be productively and meaningfully employed throughout their working careers.

Body of Management Knowledge

The manager of tomorrow should be able to profit more from the increasing body of management knowledge and experience than yesterday's and today's manager. He should be more knowledgeable about people, organizations, and the total environment. Since he will be constantly encountering change, it will be helpful if he is openminded and receptive to new ideas and new operational techniques. Patience and tolerance will be invaluable tools.

The engineer manager will need the skill and courage to make decisions himself. There will be times when he must be individually decisive. There will even be times when he must make unpopular decisions. Some of his decisions may be unpopular because they must consider long-range effects as well as short-range benefits. Other decisions may not be well received by employees because they give recognition to other publics to which the organization is responsible. The manager must, however, remember his total obligations.

The manager will receive future benefit from the ability to be objective in making decisions that affect others. Objectivity implies that an individual is capable of looking at issues and problems without allowing prejudices, biases, or emotions to obstruct reality. If the manager is objective, he is capable of viewing matters as they really are, not as he or someone else wishes they were. Being objective does not mean that the manager can have no personal feelings of his own as he works, but it does suggest that he be aware of his own feelings and keep them under control.

The work environment will be much more pleasant and more productive if the manager places an emphasis on positive motivation, on helping achieve personal goals while striving toward organization objectives. This environment induces employees to be productive because they wish to work and not because they are forced to perform. At the same time, however, the manager should not avoid the use of negative motivational techniques when they may be beneficial.

Reginald H. Jones, Chairman and Chief Executive Officer of the General Electric Company, has made the following succinct observations:

1. He will be working with more minorities and women. He will be working in an environment where equal opportunity is a way of life. More women will be qualifying themselves for leadership positions.

2. Greater political awareness. It will be necessary for him to have a greater political sophistication. Government is becoming an evermore pervasive factor in economic life. Business managers are finding the main obstacles to achieving their business objectives are external to the company.

3. More world minded. Managers of the future will be less provincial and more world minded. Today there is a struggle for world markets between multinational corporation headquarters in Europe, Japan, and North America.

4. Strategic planners. Managers of the future will be much more strategic in their thinking. Management by improvisation is out. In the future the manager will have to make careful and systematic analyses of the social, political, and economic environment in which his business must operate.

5. Greater financial sophistication. Managers of the future will be financially more sophisticated. Management of the balance sheet will become an increasingly important factor.

6. Social responsibility. Managers of the future will be much more responsive to social pressures. Economic performance is no longer enough; business is expected to act in the public interest as well as the shareowners' interest.[2]

To meet these challenges, the engineer manager of the future will have many more tools to work with than did his predecessors. Nevertheless, the challenges will be great, but so will the rewards.

[2]Reginald H. Jones, on the acceptance of the Joseph P. Horton Award, Horton Graduate School of Business, School Club of New York, March 23, 1976.

One of the authors has accumulated a number of aphorisms which may be helpful to the engineer and engineer manager.[3] They are the distillation of his 26 years of experience in a major corporation at almost all levels of management. They are passed along for whatever value they may have.

The new graduate makes his mark within the first couple of years of work within the corporation. During this period he is tagged as either a "comer" or just "one of the boys." On occasion we hear of late bloomers, but for the most part the leaders of the future make their presence known in their early days of work. This behooves the graduate to devote his maximum efforts to fulfilling the responsibilities of his job during this initial period of time. Furthermore, it never hurts to give above and beyond the call of duty.

For a maximum of success in a corporation, "somebody up there has to love you." Human beings are selfish enough that they want to look good. Therefore, when the opportunity comes to promote a man or woman, they tend to look for one who will make them look good. They will look for one in whom they have confidence and who has demonstrated his ability to perform. To develop this reputation, he must be willing to accept each opportunity as it arises and use his talents to the fullest to meet the demands of the job.

The leader or the manager is the first man on the job in the morning and the last to leave at night. General Brehon Somervell, Commander of the Army Service Forces during World War II, made this statement: "If you plan to become a leader, you will not achieve this position through adherence to a 40-hour week. You must be willing to contribute whenever and to whatever extent your services are needed."

Projects that are most likely to succeed have support from top management. It is very difficult to develop a major project or to turn the direction in which the company is going from a low-level position on the organizational totem pole. Even though there may be lip service from above, the project will undoubtedly fail. Therefore, if one cannot secure top management's active support of the project in which he is interested, it is better to set it aside.

It does no good to gripe to the man next door about your dissatisfactions. Doing this only builds up your frustrations, cuts down on the quality of your work, and enhances the possibility that you will not succeed. If you have problems or are dissatisfied about some company policy or action, go to the man who can do something about it, probably your boss. He will respect you more for it and your morale will be better.

Always do "completed staff work". When you have been given a job to do, research it thoroughly, make your decisions, and prepare a proposal. Be in a position so that you can say, "I plan to do this." If your boss doesn't like what you plan to do, he'll let you know. Don't go to him and ask, "Should I do this or

[3]Bernard R. Sarchet, Professor and Chairman of Department of Engineering Management, University of Missouri–Rolla.

should I do that?" He is paying you to make a recommendation or a decision.

Communication is two-way. Just because you understand the instructions you have given one of your subordinates does not mean that he understands them. All communications should be tested to make sure that they are understood. Have your subordinate repeat your instructions in his own words.

Do not let your problems sit on the side and hope that they will go away; they never do. If you sweep them under the rug today, you will find them there tomorrow.

Always be willing to go the other half mile. Be willing to give something extra of yourself to help your boss or a colleague.

Develop a reputation for honesty and integrity. This is extremely important in your work as an engineer or an engineer manager. You are professionally responsible for the quality of your work. It is also important in dealing with organizations, especially unions. If the engineer manager is a plant manager, it is important for him to be in a position that he can say, "Remember last year I told you thus and so, and it happened." This can be an exceedingly strong weapon in securing agreement on a contract point.

Understand the boss to the point that the decisions that you make will be the same ones that he would make. You should strive to become his alter ego.

Never let your boss be caught by surprise. If he is expecting you to deliver a report on a specific date and you won't have it ready, tell him so. It is important that he know ahead of time, because somebody is probably expecting him to make a report based on your data. Or, if something bad is happening in your plant and it is likely to come to the attention of his superiors, make sure that he knows it before they do.

Be a listening ear for your employees. Let your employees know that they can come to you with their problems. Respect and love can be secured through consideration, honesty, and support.

Don't blame your boss for policies. When you find it necessary to pass along to your employees instructions you do not necessarily favor, do not blame your boss. You are a manager and you have an obligation to act as a buffer between your boss and your subordinate. Therefore, you should not convey the impression that you were forced to pass along the decision to them. Instead, look for and give reasons why this was a good decision. If you blame your boss, you will weaken your position in the chain of command.

Develop empathy. Try to see things as your employees, your supervisor, or your peers see them. Put yourself in their skin. Walk in their moccasins.

Develop a reputation for reliability. No one wants to assign a job to an unreliable individual. When you are given an assignment, your superior is expecting a result. If he cannot depend on you to perform, he will not give you assignments in the future.

Develop your powers of judgment. No one wants to assign a job to a person who is known to have poor judgment. Strive to develop a reputation for giving consideration to all sides of an issue before making your decision.

Develop your power of enthusiasm. Nothing can be sold without enthusiasm, whether a new proposal to the management, a new decision to a group of subordinates, or a new product to the public. If you do not believe in something yourself, no one else will believe in it.

Develop your power of oral and written communication. Perhaps 90% of your time will be spent in communication with others. Unless you can communicate effectively, you will waste a good portion of your life and will reduce your chances of success. Written communication should be clear and precise.

Learn to roll with the punches because there will be many. You will encounter many disappointments during your working life. You must learn to accept them and move on.

Develop an insensitivity to criticism. If you are to be a leader, you will be criticized.

Learn to challenge and inspire your employees. The ability to do this is a mark of leadership.

Establish your goals and explain them carefully to your employees, so that they will want to follow you in achieving them.

Keep up your old school ties. Someday, your school friends may be able to help you when you are looking for a job, going on to graduate school, or looking for a recommendation letter.

Develop the ability to look beyond things as they are to what they can be. Learn to recognize that circumstances are transient. If you fail to do so, you will bog down in the mire of despair.

You will only be as successful as you think you'll be. If you do not feel successful, no one else will think you are successful.

Don't try to get four significant figure results out of an input of one significant figure data.

Never be afraid to accept other people's ideas or solicit their help. Avoid the NIH (not invented here) factor. You can greatly expand your usefulness if you will accept other people's ideas.

Learn to accept the idea that not all projects will succeed. Perhaps only one out of ten research projects started in the lab reach the marketplace. If every project were to succeed, our criteria for investigation were too narrow.

It doesn't matter how good you think you are. The only thing that counts is how good other people think you are.

If you keep these in mind as you enter the corporation or government agency, they may help to prevent you from stumbling along the way.

We have come a long way in analyzing the transition of the engineer to the engineer manager. We want to conclude the journey by providing some knowledge in the important areas of law and ethics. The three sections dealing with law cover the broad areas of the legal environment, patent law, and regulatory law. The section on ethics deals with matters that you will encounter within the corporation and that will be important to your success. Each section was written by a specialist in his field (in the law sections, by lawyers). Each one has had practical experience in industry or government and is currently teaching in his area of interest. The information should be useful to you whether you are working as an engineer or as an engineer manager.

THE LEGAL ENVIRONMENT

DAVID A. SHALLER
Assistant Professor of Engineering Management
University of Missouri–Rolla

An engineer manager has certain responsibilities and is granted certain authority as a result of his position. Within this authority, the manager has the ability to legally obligate both himself and his organization by the decisions he makes and the actions he takes. In fact, scarcely a day passes in which decisions are made and actions taken that does not involve some significant legal implications. Therefore, it is of great importance for an engineer manager to be aware of the legal environment in which he is continuously operating.

The purpose of this section is to acquaint the engineer manager with some of the introductory aspects of the legal system as it functions in this country and as it is likely to impact the manager in the performance of his job. It is impossible to cover in a few short pages everything there is to know about the legal environment; nevertheless, it is hoped that by touching on a few basic concepts of law, the reader will develop a greater ability to recognize potential legal implications of decisions and adjust his course accordingly.

For our purposes, the word "law" may be considered as that body of rules and principles which govern and control the conduct of people and organizations. Ordinarily, this body of rules is maintained and enforced in some manner by various branches of government. Yet, the outcome in any given case may be quite unpredictable as a result of the wide variety of forces which frequently come to bear upon courts in making their decisions. Thus, it would be incorrect to conceive of the law as a permanent catalog of rules which can be indexed and located in order to answer your legal questions. Invariably, there is great uncertainty and frustration in determining the precise outcome of a particular "real world" case.

Public Law and Private Law

The law may be classified into several rather general classifications. One of these is Public Law and Private Law. Public Law deals with the relationships of the government (federal, state, or local) and private individuals, or by itself. Included in this classification are Criminal Law, Administrative Law, and Constitutional Law. Each of these involves the government as the sole or major participant.

Private Law deals with the determination of the rights and duties of private individuals in relation to each other. Included in this category is the Law of Contracts, which deals primarily with agreements between individuals. Also included is Tort Law, which considers the rights and responsibilities of private individuals in their relations with each other. An example of an intentional tort would be libel or slander; an unintentional tort might be negligence. Finally, the Law of Property is included under Private Law. This area deals primarily with the rights and duties between individuals in regard to the property of others. Property Law gives rise to a complex series of relationships between persons concerning, for example, ownership.

Sources of the Law

The law is actually derived from several distinctly different sources and it is important to recognize the differences between these types of law. The touchstone of our law is Constitutional Law. Federal and state constitutions provide the highest order of written law in our system.

Statutory Law provides us with the next source of written law. Statutes, ordinances, or codes are enacted by legislative bodies such as the United States Congress or a state legislature. These statutes are written formal embodiments of the law as promulgated by the particular legislative body.

A vitally important source of the law often overlooked or unknown to many is the Common Law. Common Law is the law as stated by judges in decisions which are written in a particular case. This law is often known as case law or judge-made law. Often a judicial decision will interpret or clarify a statutory provision or establish a new legal principle in the absence of an applicable statute. Since our Common Law system is based on the principle of standing on past established precedents, it is important that the decisions be available for all to study. Almost all written case decisions or opinions are published and widely distributed and may be located in any law library. Today, much of the Common Law dealing with business is being codified into statutory form and adopted by many states. An example of this is the Uniform Commercial Code.

A fourth source of law is known as Administrative Law. It is essentially different from the previous sources in that it is neither enacted through the legislative process nor determined by a judge in exercise of his judicial duties. Yet it is likely that the majority of direct contact a business or manager will have with the law is in this area. Administrative Law is that body of rules, regulations, and procedures established by the wide variety of administrative agencies in the exercise of their quasi-legislative powers.

The Court System

Our court system provides us with two types of courts: Trial Courts and Appellate Courts. The Trial Court is sometimes referred to as an inferior or lower court, but it is the court with which most people are familiar. The Trial Court is where the case is originally filed and so it is called the court of original jurisdiction. The purpose of the trial is to present evidence bearing on a controversy. Issues of fact will be determined by a jury or by the judge in a jury-waived case. Issues involving the law as applied to the facts are always decided by the judge.

The Appellate Courts hear appeals from the Trial Courts and consider the record as submitted. This hearing is to determine whether or not any error was made by the lower court in its rulings and determinations. An appeal is not a new trial on the merits of the case; it is merely a consideration of the possibility of reversible error. The Appellate Court may affirm or reverse the case and may remand it back to the Trial Court for further proceedings. The right to an appeal is protected by requirement of Due Process.

The United States has basically two separate parallel judicial systems. The Federal Court system and the State Court system operate independently, but from time to time cases move from one system to the other. The Federal Courts include the Trial Courts, known as United States District Courts; intermediate Appellate Courts, known as United States Courts of Appeal; and the United States Supreme Court.

The District Courts are Trial Courts of limited jurisdiction, which means that only certain types of cases may be filed in them. This contrasts with courts of general jurisdiction such as state Trial Courts, in which any type of case may be filed. One of the most common forms of jurisdiction in the Federal Courts is diversity of citizenship in which the litigants are citizens of different states.

The United States Courts of Appeal have jurisdiction to hear appeals from final decisions of the District Courts as well as from many administrative agencies. The United States Supreme Court hears a very limited number of cases, and since the Due Process right of appeal is satisfied at the Circuit Court of Appeals level, there is usually no automatic right of appeal to the Supreme Court. For this reason, most cases filed are on a *writ of certiorari,* which essentially seeks to persuade the court of the importance of the particular case and the rulings of law. The important point to keep in mind is that the Supreme Court is not obligated to hear the cases filed in this manner.

State court systems typically have trial courts of general jurisdiction known as Circuit Courts or Courts of Common Pleas. In addition, most states have inferior trial courts with limited jurisdiction such as magistrate courts, small claims courts, and municipal courts. Some states have specialized courts which hear only a certain kind of case. Examples would be probate courts or family courts.

Many states, but not all, have intermediate appellate courts. Some might even have original jurisdiction, but this is unusual. Remember that Due Process requires at least one appeal as a matter of right; so in states where there is no intermediate appellate court, there must be an appeal allowed to the state's highest appellate court, usually called the Supreme Court.

Administrative agencies often perform in a judicial capacity and hold hearings which are very similar to trials. However, there is no jury and the rules are usually more relaxed. The presiding officer is a hearing officer or administrative law judge, and the appeal is to the full commission or board. From there, appeals may be taken to the established level of the state or federal judicial system. Examples of administrative agency hearings would include state workman's compensation hearings, social security appeals, and unfair labor practice hearings in the N.L.R.B.

Filing a Lawsuit

Before a court may act upon a case, it must have the power and authority, which is known as *jurisdiction*. There are two jurisdictional requirements that must be met before the court acquires its power. First, the subject matter of the case must be within the designated powers of this court. This is known as *subject matter jurisdiction*. In addition, the court must obtain jurisdiction over the person being sued. This is acquired by service of process (summons) either by personal service or as otherwise prescribed by law.

To initiate an action, a petition or complaint is filed with the court setting forth the elements of the cause of action and stating in the prayer what remedy you are seeking, e.g., damages. When service is obtained, the defendant receives a copy of the petition so that he knows the particulars of the lawsuit. Usually from 20 to 30 days are allowed to file an answer to the complaint along with a cross-complaint if any. Now the case is ready to proceed in earnest, but there are many steps before trial actually begins. Motions are often filed and heard before a judge. The discovery process takes place during this time before trial. It gives both plaintiff and defendant an opportunity to discover details about the other's case. Included here would be depositions, interrogatories, and demands for formal admissions of fact. The purpose of discovery is to allow the development of information for both sides in the hope of settlement or reducing issues at trial.

The Trial

Once a case is sent out to trial, the first order of business is the selection of a jury. The initial questioning of the panel of veniremen by the attorneys or the judge is known as the *voir dire*. This questioning is to determine that the jury selected from the veniremen will be able to give a fair and unbiased consideration to the factual issues of the case.

After the jury is empaneled, each attorney is allowed to make an opening statement in which he tells the members of the jury what the evidence is expected to show them about his case. At the conclusion of the opening statements, the plaintiff's attorney puts forth his case by introducing testimony of witnesses, exhibits, and any other evidence which may be appropriate.

At the conclusion of the plaintiff's case, the defendant's attorney may make a motion for a summary judgment or directed verdict. If he is overruled, he will then be allowed to put on his evidence in defense. Once all the evidence is before the jury, each side will be allowed to make its final arguments to the jury. These arguments are also known as *summations.* At the conclusion of the final arguments, the judge will issue specific instructions to the members of the jury which guide them in making their decisions. Sometimes these instructions are lengthy and complex. The jury is always instructed to deal with the facts only, since this is the purpose of a jury verdict. It is the judge's exclusive province in his decision to determine legal issues as opposed to factual issues. At the close of the case, a judgment is rendered determining who won the case and how much. It should be noted that winning a case and obtaining a judgment do not always mean that the winner is able to collect the money. Many a litigant has gone through his lawsuit and has successfully obtained a large judgment only to find it totally uncollectable. But that's another story!

PROTECTION OF IDEAS

DONALD D. MYERS
Assistant Professor of Engineering Management
University of Missouri–Rolla

Technical ideas may get varying degrees of protection through trade secret, patent, trademark, or copyright. The engineer manager should be familiar generally with the protection afforded by each. With this knowledge, he is in a better position to advise his subordinates and to work with the attorneys to secure appropriate protection.

Trade Secret

Coca-Cola keeps the ingredients and the mixing and brewing formulas for Coke a trade secret by locking them in an Atlanta bank vault. They are known to only two senior chemists at the company. The vault can be opened only after a special vote of the directors. There is no exact definition of a trade secret, but to be protected by the courts, trade secrets must be (1) secret, (2) substantial, and (3) valuable. The secret can be most anything so long as it is not generally known in the trade or industry to which it applies and which provides its owner with a competitive advantage. A trade secret may be a formula, process, know-how, specifications, pricing information, customer lists, sources of supply, merchandising methods, or other business information. It may or may not be protectable by other means.

The key to a trade secret is to maintain secrecy. If the secret becomes known through reverse engineering or if someone independently discovers the secret, there is no obligation to the owner of the trade secret. When it is necessary to disclose a trade secret to employees, suppliers, customers, or licensees, it may be protected through a secrecy agreement in which the recipient is obligated to guard the trade secret against loss by disclosure to others.

Patents

A utility patent granted by the United States to the inventor gives the inventor a monopoly for seventeen years to "exclude others from making, using, or selling the invention throughout the United States." A patent may be obtained for a process, machine, article of manufacture, composition of material, or any improvement thereof. Accordingly, a patent cannot be obtained on laws of nature, methods of doing business, computer software, mathematical formulas, scientific principles, or printed matter.

To be patentable, the invention must be (1) new or novel, (2) useful or have utility, and (3) nonobvious. If the invention has been used, sold, or known by others in the United States or patented or disclosed in a printed publication in the United States or a foreign country before the invention was made by the inventor, a patent is barred. It is also barred if the invention was patented or described in a publication or in public use or on sale in the United States more than one year prior to the application for the patent. Useful inventions must advance the useful arts and benefit the public. The test of obviousness is the ability of those "with ordinary skill in the art involved."

The invention process includes two steps: (1) conception and (2) reduction to practice. To the extent that the first to conceive made a reasonable diligent effort to reduce the invention to practice, he will receive the patent, even though he was not the first to reduce to practice. Accordingly, it is imperative that the inventor maintain good records to establish the date of conception and diligence in reduction to practice in case of any later interference. Reduction to practice does not always require showing perfect results. In fact, it may not require actual reduction to practice at all. The filing of the patent application satisfies reduction to practice if, from the patent specification, one skilled in the art to which it relates is capable of constructing or carrying out the invention.

Engineer managers should always conduct a patent search before proceeding with any R&D, since the eventual research may result in something that is already patented and cannot be used anyway. If the search reveals "on-point" art, the engineer manager has two choices: (1) license the technology or (2) invent around the patent. A search cannot include patent applications on file, since they are confidential. Therefore, a patent search cannot assure that no art exists.

Design patents are granted on new, original, and ornamental design of an article of manufacture for a term of three years–six months, seven years, or fourteen years as the applicant may elect. The conditions for patentability differ from utility patents, i.e., novelty, originality, and ornamentality versus novelty, utility, and nonobviousness. The design patent is not concerned with how the article of manufacture was made and what it constitutes, but how it looks.

Our patent system also permits patents for plants when asexually reproduced for cultivated sports, mutants, hybrids, and newly found seedlings other than a tuber-propagated plant or a plant found in an uncultivated state.

Patents are under the exclusive jurisdiction of the federal government. Patent applications are submitted to the United States Patent and Trademark Office.

Marks

Marks are registrable with the United States Patent and Trademark Office and include (1) trademarks, (2) service marks, (3) certification marks, and (4) collective marks. The Lanham Act defines a mark as "any word, name, symbol, or device, or any combination thereof." For federal registration, the mark must be used in interstate commerce. Registration may also be made with the various states since marks are not the exclusive jurisdiction of the federal government.

The value of a mark is generally in the goodwill that attaches to it. Accordingly, it is used to distinguish products and services to prevent others from "palming off" their products and trading on the goodwill. In many cases, the mark is of much more value than a patent, since many consumers buy based on the "recognizable name." It should be noted that a trademark differs from a trade name. The trademark attaches to a product whereas the trade name may not. IBM may be both a trademark and a trade name. It is only the trademark that is registrable with the United States Patent and Trademark Office and protected by federal statutes. If the trademark is not used in interstate commerce, it would be necessary to register it with the state for any statutory protection. The mark, versus a patent or copyright, is a creature of common law. Accordingly, registration does not add to its validity, but it is a definite benefit. Application is not made for a mark until after it has been used.

A service name is associated with services rather than goods. A certification name is used generally to indicate that goods or services with which the mark is used meet standards or requirements that have been established by the owner of the mark, e.g., Good Housekeeping. The collective name is used by members of a collective group or organization, such as membership in unions, associations, or other organizations.

It should be recognized that rights to a mark may be lost, e.g., through abandoning the mark or allowing the mark to become a generic word. Accordingly, much more vigilance is required to maintain rights to a mark.

Marks may be renewed for twenty-year periods. An affidavit must be filed within six years following the registration showing that the mark is still in use. Failure to do so will result in loss of the mark.

It is improper to indicate that a mark is registered ® prior to issuance of a certificate of registration, although it may be indicated that it is considered a mark, i.e., using ™, etc.

Copyrights

To obtain a copyright, the work must be a product of original creative authorship and fixed in some tangible form from which the work can be reproduced. The following seven broad categories are illustrative but not limitative of right to copyright protection: literary works; dramatic works; musical works, including any accompanying music; pantomimes and choreographic works; pictorial, graphic, and sculptural works; motion pictures and other audiovisual works; and sound recordings. Although not mentioned explicitly in the statutes, it is clear that computer programs or "software" is within the subject matter of copyright.

Copyright means are limited in the protection that can be afforded technical innovations. It is the form of the expression of the idea that is copyrighted rather than the idea itself. Thus, although a book may be copyrighted, giving the author the right to exclude others from copying the book, any ideas expressed, such as a description of a process, may be used by anyone unless they are protected by a patent.

Ornamental designs protectable by a design patent may also be protectable under the Copyright Act. Design patents are granted for new, original, and ornamental designs for articles of manufacture. Creative and original designs for works of art are protected by copyright. If a design may be protected by both copyright and patent, the designer must elect which will be used. The term for a design patent is three, five, seven, or fourteen years; the term for a copyright is the life of the author plus fifty years.

Labels may be trademarked or copyrighted. It is better, generally, to rely on a trademark since requirements for obtaining substantial enforcement under the copyright law for violation of a label are very strict. A label may be taken out of copyright if there is a minor change in the label although the message may remain the same.

Copyrights are the exclusive jurisdiction of the federal government. Protection by copyright is obtained by placing notice of copyright on publicly distributed copies. Although registration is not a condition of copyright protection, it is a prerequisite to an infringement suit and certain other remedies. The copyright claim is made by depositing copies together with application and fee to the copyright office in the Library of Congress.

Commercializing Technical Innovations Through Licensing

When the market for an innovation cannot be completely exploited, for whatever reason, licensing is the obvious way to maximize income. A license is a contract between the buyer and seller of the innovation. Generally, a license provides for payment through a royalty on sales of products incorporating the innovation.

The license agreement itself can be as flexible generally as the parties contracting may agree. It may increase use of the innovation for making, using, and selling, or any one or combination of the three functions. Making, using, or selling may be limited to a specific territory, product, or market. The agreement may give exclusive rights or the seller may retain the right for his own use and/or the right to license others.

Arguments may be made for licensing competition. Many purchasers of goods require at least two sources. Accordingly, if the supplier has some innovative product that he wants to sell, it would be necessary to license others. The innovator still maintains a competitive advantage since his costs do not include the royalty being collected from his competitor. Frequently industries standardize on certain items or processes. Once they make the selection of a standard, they are reluctant to change. Accordingly, to get the innovation selected as a standard, it may be necessary to license others to make the item readily available for general acceptance as a standard.

The engineer manager should always consider licensing other's technology as an alternative to internal development. Internal development may be more costly and result in a critical loss of time.

Congress is showing increased interest in motivating government agencies, industry, and universities in transferring technologies. It is generally agreed that the slowdown in the United States' productivity is a major contributor to the weakening economy. By improving the availability of innovative technology, productivity will be improved. Engineer managers of the future will need to recognize the significance of maximizing their resources through licensing of technology.

Managing Technical Innovations

Technical innovation is significant in our lives on a macro as well as a micro scale. On a macro level, it affords the most likely means for improving productivity of the United States' economy. It is generally recognized that the United States has been dilatory in its technical innovation during the 1970s. This is evident when the productivity of other industrial nations is compared to that of the United States. Further, evidence is provided when the statistics of the United States Patent and Trademark Office are reviewed.

On a micro scale, when we consider the individual or a specific company, technical creativity is extremely significant. The American dream of making it big through technical innovation has occurred over and over in American history. Henry Ford's Model A changed the nation. Production of vehicles increased from 1.9 million in 1919 to 5.6 million in 1929. As a result, construction of thousands of miles of roads was spurred. Lining these roads were garages, filling stations, hot dog stands, tourist stops, and campsites. The oil industry expanded enormously, as well as rubber, steel, and other Detroit-related businesses. Today, we see similar effects with the coming of age of the computer industry.

How to Establish Environment for Technical Innovation

The engineer manager, being responsible for managing creative ideas, must address the issue of how this important responsibility will be accomplished. We must acknowledge that the individual engineer manager is limited substantially by the environment provided. However, it is important that he recognize how the significance of the external and internal environment created for the employees impacts the degree of innovation.

Government attitudes toward technical innovations are reflected in the percentage of its budget directed toward research and development. They can also be seen in the tax policies adopted relative to the treatment of depreciation and monies expended for R&D. Government contracting policies with the private sector also impact R&D write-off by industry as well as patent retention rights. Foreign sales may be severely limited by government export regulations. Attitudes of the courts and the United States Patent and Trademark Office also affect technical innovation. The individual engineer manager can probably influence these policies and attitudes most effectively by working through industry and professional associations as well as legislators.

At the company level, attitudes and policies have a significant impact on technical innovation. The engineer manager has considerably more influence on these matters. It is important that this significance be communicated to top management. Policies for the role of R&D must be adopted, i.e., targets for improvements, depth of research effort, offensive versus defensive R&D, using outsiders for R&D, and limits on total commitments. Company policies toward sharing the rewards of technical innovations also influence creativity.

The engineer manager's largest impact on the technical innovation process will be through the attitudes and policies he adopts for his department in implementing company policies. It is particularly important to recognize the need for creating a working environment that is conducive to innovation.

ECOLOGY AND THE LAW

GORDON WEISS
Instructor of Engineering Management
University of Missouri–Rolla

William H. Rodgers, Jr., in the introduction to his compendium on environmental law, states that "Discovering a workable definition of environmental law is a little bit like the search for truth: the closer you get, the more elusive it becomes."[4] Most people in management today might agree that the definition of environmental law is somewhat vaporous; however, they would be certain to add that its impact has been anything but elusive and they would probably hasten to

[4]William H. Rodgers, Jr., *Environmental Law* (St. Paul, Minn.: West Publishing Co., 1977), p. I.

acknowledge that their decision-making procedures have been altered by it to a most remarkable degree.

Thomas F. P. Sullivan, the president of Government Institutes Inc., points out in his handbook on environmental law that this segment of the law "encompasses all the protections for our environment that emanate from the U.S. Constitution, the state constitutions, the federal and state statutes and local ordinances, regulations promulgated by federal, state, and local regulatory agencies, court decisions interpreting these laws and regulations, and certain doctrines of the Common Law."[5]

To generalize, environmental law concerns itself with optimization, preservation, and control of the environment. Prevention of practices which result in pollution of the environment is the central theme of legislation and regulations dealing with the subject.

Although we are prone at times to look upon environmental laws and concern for the ecology as a hallmark of our time, such is not really the case.[6] The negative effects to society of numerous forms of pollution were recognized many years ago in this country and an ancient English case recognized the offensiveness of a hog sty and granted an injunction against it centuries ago.[7]

In this country there were a number of notable pioneers in this field just prior to and after the turn of this century; among these individuals who had great concern for environmental matters were Theodore Roosevelt, Gifford Pinchot, George Perkins Marsh, and John Muir, all of whom worked diligently for wildlife conservation, preservation of scenic natural areas, and controlled exploitation of natural resources.[8] There were a number of federal legislative efforts directed at preservation, enhancement, and control of the environment between 1899 and 1970, but for the most part, these called primarily for the catalyzation of action via conscience. For the most part, controls over deleterious practices which damage the environment were left to the states to correct. Individual controversies were usually settled in court by application of the common law doctrines of nuisance, negligence, trespass, and strict liability. But there was no significant national recognition of the major pollution problems that were building up in this country by the public sector in general.

National concern for the environment began sweeping the country in the 1960s and culminated in the passage of the National Environmental Policy Act of 1969 (42 USC §4321). This was followed by feverish Congressional efforts to amend environmental laws that had been comparatively weak and to pass new, broad, sweeping omnibus legislation that addressed specific problems relative to environmental matters. This proliferation of federal legislation resulted in similar statutory and regulatory expansions at the state and local levels. Today, there is a broad spectrum of federal statutory law dealing with the environment, as follows:

[5]Thomas F. P. Sullivan, *Environmental Law Handbook* (Washington, D. C.: Government Institute, Inc., 1979), p. 2.

[6]61 American Jurisprudence 2^d, §2.

[7]Alfreds Case, 77 English Reports, 816 (KB) Kings Bench.

[8]61 American Jurisprudence 2^d, §2.

1. *The Clean Air Act* (42 USC §1857 et seq.): The purpose expressed in this Act is "to protect and enhance the quality of the Nation's air resources so as to promote the public health and welfare and the productive capacity of its population."

2. *The Federal Water Pollution Control Act,* as amended 1972 (33 USC §1251 et. seq.): The objective of this Act as expressed in Sec. 101(a) is "to restore and maintain the chemical, physical, and biological integrity of the Nation's waters."

3. *The National Environmental Policy Act of 1969* (42 USC §4321): The purpose of this Act as set down in Sec. 2 is as follows: "to declare a national policy which will encourage productive and enjoyable harmony between man and his environment; to promote efforts which will prevent or eliminate damage to the environment and biosphere and stimulate the health and welfare of man; to enrich the understanding of the ecological systems and natural resources important to the Nation; and to establish a Council on Environmental Quality."

4. *The Noise Control Act of 1972* (42 USC §4901): In this Act, the Congress declared in Sec. 2(b) that "it is the policy of the United States to promote an environment for all Americans free from noise that jeopardizes their health or welfare. To that end, it is the purpose of this Act to establish a means for effective coordination of federal research activities in noise control, to authorize the establishment of federal noise emission standards for products distributed in commerce, and to provide information to the public respecting the noise emission and noise reduction characteristics of such products."

5. *The Occupational Safety and Health Act* (29 USC §615): Sec. 2(b) of this Act states: "The Congress declares it to be its purpose and policy, through the exercise of its powers to regulate commerce among the several states and with foreign nations and to provide for the general welfare, to assure so far as possible every working man and woman in the Nation safe and healthful working conditions and to preserve our human resources."

6. *The Toxic Substances Control Act* (15 USC §2601 et seq.): Sec. 2(b) of this Act states: "It is the policy of the United States that:

    A. Adequate data should be developed with respect to the effect of chemical substances and mixtures on health and the environment and that the development of such data should be the responsibility of those who manufacture and those who process such chemical substances and mixtures;

    B. Adequate authority should exist to regulate chemical substances and mixtures which present an unreasonable risk of injury to health or the environment, and to take action with respect to chemical substances and mixtures which are imminent hazards; and

    C. Authority over chemical substances and mixtures should be exercised in such a manner as not to impede unduly or create unnecessary economic barriers to technological innovation while fulfilling the primary purpose of this Act to assure that such innovation and commerce in such chemical substances and mixtures do not present an unreasonable risk of injury to health or the environment."

7. *The Resource Conservation and Recovery Act of 1976* (42 USC §6901 et seq.): In Sec. 1003 of this federal Act, Congress has spelled out its objectives as follows:

"A. Providing technical and financial assistance to state and local governments and interstate agencies for the development of solid waste management plans (including resource recovery and resource conservation systems) which will promote improved solid waste management techniques (including more effective organizational arrangements), new and improved methods of collection, separation, and recovery of solid waste, and the environmentally safe disposal of nonrecoverable residues;

B. Providing training grants in occupations involving the design, operation, and maintenance of solid waste disposal systems;

C. Prohibiting future open dumping on the land and requiring the conversion of existing open dumps to facilities which do not pose a danger to the environment or to health;

D. Regulating the treatment, storage, transportation, and disposal of hazardous wastes which have adverse effects on health and the environment;

E. Providing for the promulgation of guidelines for solid waste collection, transport, separation, recovery, and disposal practices and systems;

G. Promoting a national research and development program for improved solid waste management and resource conservation techniques, more effective organizational arrangements, and new and improved methods of collection, separation, and recovery, and recycling of solid wastes and environmentally safe disposal of nonrecoverable residues;

H. Promoting the demonstration, construction, and application of solid waste management, resource recovery, and resource conservation systems which preserve and enhance the quality of air, water, and land resources; and

I. Establishing a cooperative effort among the federal, state, and local governments and private enterprise in order to recover valuable materials and energy from solid waste."

The foregoing material is representative of the emphasis the Congress has placed on environmental concerns during the past decade. Supplementary to these "ecology-focused" measures are legislative mandates and judicial decisions dealing with sovereign immunity, the gamut of administrative law, freedom of information, citizen suits, population growth, judicial review of agency actions, and scientific and technological assessments. In addition, issues between states bring the federal common law into the picture. Thus, it becomes very difficult to set rigid parameters on exactly what areas of the law actually impact questions and issues affecting the environment.

What caused this proliferation of environmental legislation? There is no simple answer to that question because the environmental problems are so complex because there is enormous synergy between them. Among the more general contributing factors cited by leading environmental law writers are the following:

1. Since the United States has always enjoyed good balance among the major factors of production, we have emphasized quantitative growth and tended to ignore qualitative considerations in regard to the way we treated our natural surroundings.

2. Sociologists and environmentalists charge that we failed in the past to attach cost/benefit values to our corruption of the environment and that we have neglected to consider these factors in our long-range planning efforts.

3. It has been said that many of our governmental institutions had tunnel vision in regard to environmental concerns and were actually guilty of expanding environmental problems while carrying out their individual programs.

4. We, as a nation endowed with enormous wealth, became obsessed with conveniences which were actually detrimental to the environment.

5. We have failed to look upon the environment as a total system and to understand that every subsystem and every element within it has an ultimate impact on our ecological position and profile.[9]

The passage of broad spectrum federal legislation has resulted in the delegation of implementing responsibilities to numerous federal agencies. The one primarily responsible for implementation and enforcement of the federal environmental statutes is the Environmental Protection Agency. But other federal agencies have specific responsibilities in protecting the environment and all of them are responsible for compliance with the mandates inherent in the National Environmental Policy Act.

In addition, Congress has encouraged the individual states to structure enforcement and monitoring agencies and to pass environmental legislation necessary to their function. It is a well-accepted truism that in order for federal environmental policy and legislation to have optimal impact, the state must develop a highly efficient organizational structure for dealing with environmental problems at the regional and local levels. In essence, the federal government has spelled out a strategic plan for optimizing the environment and the states are expected to develop operational programs sufficient to its implementation. And this brings up a matter of crucial importance and one of major significance to those who are involved in engineering management, and it seems most appropriate to include this matter in this capsulized presentation.

First of all, the impact of statutory and regulatory mandates occurs in two

---

[9]Jerome G. Rose, ed., *Legal Foundations of Environmental Planning* (New Brunswick, N.J.: Rutgers University, 1974).

distinctive steps. The initial or primary impact of environmental legislation at the federal level has been institutional. This means that it resulted, at the outset, in the creation, alteration, and adjustment of agencies which were delegated authority and responsibility, at both the state and federal level, to implement the mandates inherent in the environmental statutes. Thus, the initial impact of the environmental laws was not long in being felt. The secondary impact (that of direct effect on the environmental problems addressed) may take years. Why? Because it is necessary for the implementing agencies to develop levels of operating efficiencies never before contemplated in order to successfully apply the statutes.

No policy, statute, or regulation can implement itself. It requires that an organization devote itself to its implementation. Many agencies, especially at the state level, are required to analyze the parameters of a given environmental problem and develop a strategy to deal with it. In most cases, they must plan broad-based alternative strategies. Then they must develop specific program packages for carrying out all aspects of correcting the overall environmental problem assigned to them. They must formulate and publish regulations and standards within the constraints of highly complex technical areas. They must develop innovative information systems that will aid them in delineating problems and in subsequently monitoring them. In addition, they must appraise technological developments in light of environmental impacts and adjust their operational procedures in accordance with political fluctuations, manpower limitations, and budgetary inadequacies.

In short, if environmental laws and regulations are to realize their maximum impact on the problems they address, then the implementing agencies are, of necessity, going to be required to constantly upgrade their operational management skills. Since they are working in areas that require extensive technical capability, they will need to utilize engineer managers to the extent possible.

There is no dearth of environmental laws and regulations. But for the ultimate positive effect they have will depend on the level of skill achieved by their implementers in planning, organizing, staffing, appraising, and controlling.

MANAGERIAL ETHICS

ROBERT S. BAREFIELD
Associate Professor of Engineering Management
University of Missouri–Rolla

The most influential ethical philosopher, Immanuel Kant, in the conclusion of his *Critique of Practical Reason* testified to his belief in the inherent reality of morality. He wrote: "Two things fill the mind with ever-increasing admiration and awe, the oftener and the more steadily we reflect on them: the starry heavens above and the moral law within."[10]

[10]Immanuel Kant, *The Critique of Practical Reason*, Vol. 42 of Great Books of the Western World (Chicago: Encyclopaedia Britannica, Inc., 1974).

The study of morals and ethics is usually divided into two aspects, descriptive and normative ethics. Descriptive ethics is the empirical study of the moral beliefs and practices of various people and cultures. It seeks to understand the influence of social, geographic, and economic factors in these moral practices. The social sciences are involved in the study of descriptive ethics.

Normative ethics is not concerned with describing existing ethical practices, but rather with questioning the existing social morality. It inquires about the nature of good and evil, with what we should or ought to do.

Two major influences from normative ethics have had a strong influence on business ethics: ethical egotism and the normative ethics of Immanuel Kant.

Ethical egotism's influence within the business community was mediated in America primarily by Adam Smith. He believed that the public good often evolves out of competing self-interests. It is as if an "invisible hand" translates self-interest into social good. The competitive market itself would naturally work toward the best interests of the larger community. Government was to be limited, with its main function being to ensure that competition turned self-interest into public good. Historically, the business world has liked this egotistic view of economic society. A rationale for self-interest is justified on utilitarian grounds.

Kant's emphasis on an inherent moral law militates against obvious unethical practices. He suggests the concept of universality as a guide to measure ethical practices. If cheating, bribes, and/or kickbacks were universally practiced, our social structure would be damaged, perhaps destroyed. Such actions are universally immoral, quite independent of the specific circumstances. Also, such rules as the sanctity of contracts and the notion of "my station and its duties" rely on the normative concept of universal morality.

We shall now consider the following aspects of managerial ethics: (1) corporate social responsibility, (2) conflicts of interest, and (3) advertising.

Corporate Social Responsibility

There is a continuing debate in America today concerning this topic. The extreme conservative position, as exemplified by economist Milton Friedman, argues that "there is one and only one social responsibility of business—to use its resources and engage in activities designed to increase its profit so long as it stays within the rules of the game, which is to say, engages in open and free competition, without deception or fraud."[11] Friedman depends heavily on the ethical egotism of Adam Smith to support his position. Thus, he believes that we should still be guided by the ethical considerations espoused by Smith, whose basic book, *The Wealth of Nations,* was published in 1776.

A more liberal position holds that business, like an individual, needs to act responsibly regarding the consequences of its actions. Thus, a socially responsible business will act in such a way that it will improve the social quality of life as

[11]Milton Friedman, *Capitalism and Freedom* (Chicago: University of Chicago Press, 1962).

well as its own profits. Since World War II American society has placed pressure on the business world to be socially responsible, and intelligent businessmen have accepted a corporate social responsibility.

A basic aspect of social responsibility pertains to the environment. It is now generally accepted that businesses should not be concerned narrowly with the production and assumption of goods and services, but broadly with the quality of life. The pollution of the environment has become a legal as well as a moral issue. This is well illustrated by the $13.2 million fine that Judge Robert Merhige levied against Allied Chemical in the well-publicized *Kepone* case. In his opening statement of the sentencing, Judge Merhige said that the environment belongs to every person in the country. Thus, by implication, those who harm the environment are liable to society for the damage done.

Affirmative action hiring has become in recent years another example of social responsibility that has both moral and legal aspects. Government policies and laws in this regard are intended to ensure fairer opportunities for women and minority groups. An effort is made to compensate for past discrimination against persons on the basis of race, sex, nationality, or religion. "Preferential hiring" to meet affirmative action goals is the means usually used to implement affirmative action programs. This gives preference in recruitment to women, minority groups, and others discriminated against in the past. One basic concern with affirmative action programs is that they will lead to "reverse discrimination," i.e., that preferential hiring to redress previous injustices will lead to discrimination against other previously favored groups.

Conflicts of Interest

Conflicts of interest are a major concern of business and managerial ethics. A conflict of interest results when two interests claim our attention—usually one is personal and the other professional—and the implementation of one interest violates a moral, ethical, or legal principle. Thus, a conflict of interest results when accepted business principles, codes, or laws are interpreted to further the self-interest of the individual.

The celebrated Bert Lance case is a recent example. He was a close personal friend of President Jimmy Carter. His conflict of interest arose primarily from his using his position as president of a Georgia bank to obtain privileges that ordinary citizens do not have. His personal bank account was frequently heavily overdrawn. Yet, he was not penalized. Also, he received large personal loans from other banks because of deposits made from his bank. Thus, the bank's influence was used by Lance for personal gain.

Advertising

Advertising is a concern of the business world that has many ethical and moral implications. Advertising is important and necessary to the life of the nation. Yet it is a function that can be easily misused and corrupted. Many cases are taken to the courts by the Federal Trade Commission, which has the responsibility of regulating advertising. Advertising can use misleading statements, untrue claims, phony testimonials, and deceptive language to distort the truth in order to sell its products. Also, the use of sexually enticing females in the advertising of a variety of products is questionable from an ethical viewpoint. The misuse of advertising is so flagrant that elementary-age school children can discriminate between valid and false or misleading statements. The writer is aware of a unit on advertising that successfully accomplishes this ability to discriminate at the fifth-grade level. Although advertising may well be the most difficult area of corporate social responsibilities in which to maintain high standards, the corporate management bears the responsibility to ensure that its company advertising is both ethical and legal.

CLOSING REMARKS   Our journey through the transition of the engineer to engineer manager has come to an end. We have attempted to relate matters that you are acquainted with as an engineer to your new position as an engineer manager. We have provided some tools that we feel will be useful to you in your performance as a manager. In this chapter we took a look at engineering management in the future, provided a few aphorisms which may be helpful to your success, and included some additional important information in the areas of law and ethics.

So in closing, as you are successful as engineer managers, to that extent will this book have been a success and will our society have profited by it. Our future and the future of those who follow us will be influenced by your performance as an engineer manager.

Important Terms

Common Law: Body of principles and rules of action which derive their authority from judgments and decrees of courts as distinguished from statutes and constitutional law.

Copyright: The sole right of a writer or artist to reproduce, publish, and sell a literary or artistic work.

Descriptive Ethics: The empirical study of the moral beliefs and practices of various peoples and cultures.

Ethical Egotism:	The concept that the public good evolves out of competing self-interests.
License Agreement:	A contract granting rights to others to a legally protected innovation.
Mark:	Any word, name, symbol, or device used to identify products or services, provide certification of products or services, or indicate membership in an organization.
Normative Ethics:	The study of existing social morality. It inquires about the nature of good and evil and with what we should or ought to do.
Patent:	The sole right of an inventor to exclude others from making, using, or selling his invention.
Policy:	The general principles by which a government is guided in its management of public affairs or by which the legislature is guided in its measures.
Regulation:	A rule or law prescribed by a governmental agency by which certain conduct or actions are controlled.
Statute:	A law enacted and established by the will of the legislative department of government.
Technical Innovation:	The discovery or invention of a new idea, method, or device.
Tort:	A private or civil wrong or injury. A wrong independent of contract.
Trade Secret:	A secret in a trade used to obtain an advantage over competition such as a formula, pattern, device, or business information.
Venireman:	A member of a panel of prospective jurors.
Voir Dire:	The preliminary examination the court makes of a person presented as a witness or juror to determine his competency, interest in the case, etc.

For Discussion

1. Explain the dual system of courts in this country.
2. Distinguish between the role of a trial judge and that of a reviewing justice in an appellate court.
3. Explain the significance of the following statement: "Federal Courts are courts of limited jurisdiction."
4. What are the means available for protection of technical innovations?
5. How is the United States economy improved by technical innovation?
6. Why has concern for the environment been the subject of so much attention by the U.S. Congress over the past decade?
7. What role does a government agency play in causing a federal statute to have impact on an environmental problem?
8. Why is it difficult to define environmental law clearly and concisely?
9. Do you agree with Milton Friedman that the primary social responsibility of business is to maximize profits?
10. How can advertising be made more socially responsible?
11. Do you support "preferential hiring" to meet affirmative action goals?

For Further Reading

BEAUCHAMP, TOM L., and NORMAN E. BOWIE, *Ethical Theory and Business.* Englewood Cliffs, N.J.: Prentice-Hall, Inc., 1979.

CORLEY, ROBERT N., ROBERT L. BLACK, and O. LEE READ, *The Legal Environment of Business,* 4th ed. New York: McGraw-Hill Book Company, 1977.

BURGE, DAVID A., *Patent and Trademark Tactics and Practice.* New York: John Wiley & Sons, Inc., 1979.

DONALDSON, THOMAS, and PATRICIA H. WERHARE, *Ethical Issues in Business.* Englewood Cliffs, N.J.: Prentice-Hall, Inc., 1979.

FINNEGAN, MARCUS B., and ROBERT GOLDSCHNEIDER, *The Law and Business of Licensing* (4 vols.). London: Clark Boardman Co., Ltd., 1978.

GOLDSTEIN, PAUL, *Copyright, Patent, Trademark and Related State Doctrines.* Chicago: Callaghan and Company, 1973.

KINTNER, EARL W., and JACK L. LAHR, *An Intellectual Property Law Primer.* New York: Macmillan Publishing Co., Inc., 1975.

LIEBERSTEIN, STANLEY H., *Who Owns What Is in Your Head? Trade Secrets and the Mobile Employee.* New York: Hawthorn Books, Inc., 1979.

LITKA, MICHAEL P., and JAMES E. INMAN, *The Legal Environment of Business: Text Cases and Readings,* 2nd ed. New York: Grid Publishing Co., 1979.

MISSHAUK, MICHAEL J., *Management.* Boston: Little, Brown & Company, 1979.

SCHANTZ, WILLIAM T., *The American Legal Environment.* St. Paul, Minn.: West Publishing Co., 1976.

VAUGHN, RICHARD C., *Legal Aspects of Engineering,* 3rd ed. New York: Kendall/Hunt Publishing Company, 1977.

INDEXES

*Authors in this index appear in footnotes on page indicated

SUBJECT INDEX